Be prepared...
To learn...
To succeed...

Get **REA**dy. It all starts here.
REA's preparation for the GHSGT in Mathematics is **fully aligned** with the Quality Core Curriculum adopted by the Georgia Department of Education.

Visit us online at
www.rea.com

Ready, Set, Go!™

GHSGT
Mathematics

2nd Edition

Staff of Research & Education Association

Research & Education Association

The Quality Core Curriculum in this book was created and implemented by the Georgia State Board of Education. For further information, visit the Board of Education website at *http://public.doe.k12.ga.us*.

Research & Education Association
61 Ethel Road West
Piscataway, New Jersey 08854
E-mail: info@rea.com

Ready, Set, Go!
Georgia GHSGT Mathematics Test

Copyright © 2009 by Research & Education Association, Inc.
Prior edition copyright © 2007 by Research & Education Association, Inc. All rights reserved. No part of this book may be reproduced in any form without permission of the publisher.

Printed in the United States of America

Library of Congress Control Number 2008925160

ISBN-13: 978-0-7386-0443-5
ISBN-10: 0-7386-0443-7

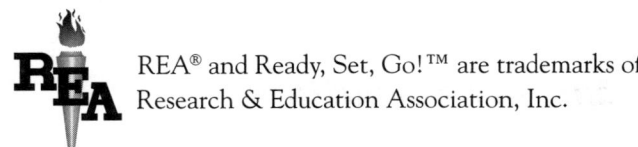

REA® and Ready, Set, Go!™ are trademarks of Research & Education Association, Inc.

Contents

Introduction 1

Chapter 1: Numbers and Computation, Part 1 11
    Standards 11
    Equivalent Numbers 12
Let's Review 1: Equivalent Numbers 20
        Patterns 22
Let's Review 2: Equivalent Expressions 27
Chapter 1 Review 29
Chapter 1 Answers 33

Chapter 2: Numbers and Computation, Part 2 37
    Standards 37
    Estimation 38
Let's Review 3: Estimation 43
        Computing Money 44
Let's Review 4: Computing Money 47
Chapter 2 Review 49
Chapter 2 Answers 52

Chapter 3: Data Analysis, Part 1 55
    Standards 55
    Probability 55
Let's Review 5: Probability 58
        Mean, Median, Mode, and Range 61
Let's Review 6: Mean, Median, Mode, and Range 63
Chapter 3 Review 66
Chapter 3 Answers 70

Chapter 4: Data Analysis, Part 2 75
    Standards 75
    Pictographs 75

Bar Graphs	76
Line Graphs	77
Circle Graphs	78
Venn Diagrams	78
Let's Review 7: Graphs and Venn Diagrams	79
Chapter 4 Review	82
Chapter 4 Answers	86

Chapter 5: Measurement and Geometry, Part 1 89

Standards	89
Customary Measures	89
Let's Review 8: Measurement	91
Metric Measures	92
Let's Review 9: Metric Measurement	94
Time	95
Area	96
Volume	97
Let's Review 10: Time, Area, and Volume	98
Chapter 5 Review	100
Chapter 5 Answers	103

Chapter 6: Measurement and Geometry, Part 2 107

Standards	107
Congruent Figures	108
Similar Figures	108
Let's Review 11: Figure Measurement	110
Transformations	111
Let's Review 12: Figure Transformations	112
The Coordinate Plane	113
Let's Review 13: Coordinate Plane	116
Chapter 6 Review	118
Chapter 6 Answers	122

Contents

Chapter 7: Measurement and Geometry, Part 3 125
- Standards 125
- Perimeter 126
Let's Review 14: Perimeter 129
- Lines 131
- Angles 132
Let's Review 15: Lines and Angles 134
- Triangles 134
- Circles 137
Let's Review 16: Triangles and Circles 138
Chapter 7 Review 139
Chapter 7 Answers 141

Chapter 8: Algebra, Part 1 143
- Standards 143
- Simplifying Expressions 143
Let's Review 17: Simplifying Expressions 145
- Evaluating Expressions 146
Let's Review 18: Evaluating Expressions 147
- Solving Equations 148
- Translating Words in Expressions and Equations 149
- Solving Problems with Formulas 149
Let's Review 19: Solving Equations 150
Chapter 8 Review 152
Chapter 8 Answers 155

Chapter 9: Algebra, Part 2 159
- Standards 159
- Proportions 160
Let's Review 20: Proportions 161
- Linear Inequalities 162
- Linear Equations 163

Let's Review 21: Linear Inequalities and Linear Equations **165**
       Determining Slope 167
Let's Review 22: Determining Slope **168**
Chapter 9 Review **170**
Chapter 9 Answers **173**

Practice Test 1 177

Practice Test 1 Answers 197

Practice Test 2 205

Practice Test 2 Answers 225

Index 231

About Research & Education Association

Founded in 1959, Research & Education Association is dedicated to publishing the finest and most effective educational materials—including software, study guides, and test preps—for students in middle school, high school, college, graduate school, and beyond. Today, REA's wide-ranging catalog is a leading resource for teachers, students, and professionals. We invite you to visit us at www.rea.com to find out how REA is making the world smarter.

Acknowledgments

    We would like to thank REA's Larry B. Kling, Vice President, Editorial, for supervising development; Pam Weston, Vice President, Publishing, for setting the quality standards for production integrity and managing the publication to completion; Michael Reynolds, Managing Editor, for project management and preflight editorial review; Christine Saul, Senior Graphic Artist, for cover design; and Jeff LoBalbo, Senior Graphic Artist, for post-production file mapping.

    We also gratefully acknowledge the writers, educators, and editors of REA for content development and Caragraphics for page design and typesetting.

Introduction

Passing the GHSGT Mathematics Test

About This Book

This book will provide you with an accurate and complete representation of the Georgia High School Graduation Test (GHSGT) in Mathematics. Inside you will find targeted reviews that are designed to provide you with the information and strategies needed to do well on these tests. You'll also find two full-length practice tests based on the official GHSGT. These practice tests contain every type of question that you can expect to encounter on the GHSGT. Following each test, you will find an answer key with detailed explanations designed to help you completely understand the test material.

About the Test

Who Takes These Tests and What Are They Used For?

The GHSGT is given to all students throughout Georgia who have entered the ninth grade since July 1, 1991. It is given to ensure that graduating students have mastered essential core academic content and skills. The test is given in four content areas: Mathematics, English Language Arts, Science, and Social Studies.

The GHSGT measures achievement in the skills and competencies outlined in the Georgia Quality Core Curriculum (QCC). Students must pass the test in order to earn a high school diploma; however, students who do not pass the test the first time are given many retest opportunities so that they may graduate in the spring of their twelfth-grade year. Those who fail to pass the test at this point, but have met all the other requirements necessary for graduation, may be able to obtain a Certificate of Performance or a Special Education Diploma. Students who leave school with either of these documents may retest again as often as necessary to obtain a high school diploma.

Is There a Registration Fee?

No.

When and Where Is the Test Given?

The GHSGT is administered to Georgia high school students for the first time in Grade 11. The GHSGT is administered in the spring, with retest opportunities in the following fall. Students are given five opportunities to take the GHSGT in each content area before the end of their twelfth-grade year.

Test Accommodations and Special Situations

Every effort is made to provide a level playing field for students with disabilities taking the GHSGT and seeking a standard high school diploma. Waivers and variances are granted for students who meet certain criteria.

A waiver is a decision by the State Board of Education (SBOE) not to apply all or part of the requirements of the GHSGT to a Georgia student who meets certain basic qualifications, such as:

- A disability, documented in a student's Individual Education Program (IEP), that makes the student incapable of passing a section of the test, even with specified testing accommodations.
- A substantial hardship beyond the student's control that has prohibited the student from having a reasonable opportunity to pass a section of the GHSGT.

An accommodation is an adjustment that is made to the testing situation based on a disability and identified in a student's IEP. Accommodations may include adjustments in the test setting, the amount of time provided in which to take the test, the way in which the test is administered, or the need for assistive technology.

A variance is a decision by the State Board of Education (SBOE) to modify all or part of the literal requirements for the GHSGT for students who have

- attempted the relevant section(s) of the GHSGT four or more times without passing and the most recent attempt is within the last calendar year; and

- successfully completed a structured remedial class(es) after each required attempt to pass the relevant section(s) of the GHSGHT; and

- passed any three of the graduation tests (four content sections of the GHSGT); and

- met the attendance and course requirements for graduation defined by the SBOE for the student's graduating class; and if the students has a 90 percent or better attendance record, excluding excused absences, while enrolled in grades 9–12; and

- at any time obtained a scaled score that falls within one standard error of measurement (SEM) for passing the relevant section of the GHSGT; and

- successfully passed each of the End-of-Course Tests (EOCT) related to the sections of the GHSGT in which the variance is being sought.

Federal law requires that students with disabilities must participate in statewide assessments such as the GHSGT. Students seeking a waiver or variance must request consideration for a waiver through their local superintendent. More information on variances and waivers may be obtained at the Georgia Department of Education website at *www.doe.k12.ga.us*. Additional resources to help you prepare to take the GHSGT may also be found on this website. Students may also ask questions of their school counselors.

How to Use This Book

What Do I Study First?

Read over the review sections and the suggestions for test taking. Studying the review sections thoroughly will reinforce the basic skills you need to do well on the test. Be sure to take the practice tests to become familiar with the format and procedures involved with taking the actual GHSGT.

When Should I Start Studying?

It is never too early to start studying for the GHSGT—the earlier you begin, the more time you will have to sharpen your skills. Do not procrastinate! Cramming is *not* an effective way to study, since it does not allow you the time needed to learn the test material. The sooner you learn the format of the exam, the more time you will have to familiarize yourself with the exam content.

Overview of the GHSGT

The sixty multiple-choice questions on the mathematics portion of the GHSGT are based on four broad strands, which embrace a series of standards:

Strand 1 standards focus on Number and Computation.

Strand 2 standards focus on Data Analysis.

Strand 3 standards focus on Measurement and Geometry.

Strand 4 standards focus on Algebra.

ITEM TYPES

Number and Computation, Data Analysis, Measurement and Geometry, Algebra

Item Type	Stimulus Characteristics	Cognitive Level	Correct Response Characteristics
1	Direct question requiring recall of facts and definitions	Low	Demonstrates knowledge of facts and basic ideas
2	Direct question requiring some interpretation or simple computation (one-step problem)	Medium	Demonstrates ability to substitute values in formulas and equations; identifies appropriate operation, unit of measure, type of graph, or geometric figure; applies problem-solving skills to real-word situations
3	Direct questioning requiring application of mathematical theories, analysis, or more complicated problem situations (two- and three-step problems), evaluating data and drawing conclusions	High	Demonstrates the ability to solve complex problems, analyze data, apply mathematical principles to real-world situations, differentiate between correct and incorrect responses

Cognitive levels are based on learning expectations, not item difficulty. However, cognitively higher-level test items generally prove to be more difficult. Here's a breakdown of the three basic cognitive levels.

Low: Requires recognition only and typically deals with terminology, identification, or other low-level activities

Medium: Requires some degree of interpretation of a problem or situation in which a mathematical principle is applied

High: Requires a significant degree of interpretation, problem solving, and analysis (e.g., devising a solution to a problem by applying a mathematical principle)

Summary of Mathematics Test Content

The following are standards from the Quality Core Curriculum in mathematics.

Strand 1: Number and Computation (17–19% of the test), Chapters 1 and 2

1. Expresses numbers in equivalent and approximate forms and orders these forms, using appropriate tools such as calculators (includes fractions, decimals, percent; scientific notation; square and cube roots, and second and third powers of whole numbers; approximations of fractions, decimals, and percents).

2. Recognizes, describes, and applies certain patterns for addition and multiplication.

3. Selects and uses problem-solving strategies and computational tools (mental computation, calculator, estimation, paper and pencil) to solve simple problems involving career, consumer, and leisure applications, and evaluates reasonableness of results.

4. Determines amounts of money, including price, amounts of change, discounts, sales prices, sales tax, interest, and best buy.

5. Uses estimation strategies such as rounding, front-end estimation, clustering, grouping, adjusting, compensation, and reference point to predict computational results.

6. Uses estimation and approximation to check the reasonableness of computational results.

7. Recognizes appropriate practical situations in which to use and to expect results with exact and approximate numbers.

Strand 2: Data Analysis (19–21% of the test), Chapters 3 and 4

8. Uses probability correctly to predict outcomes of given events, determines the probability of an event through experiments, and differentiates odds from probability.

9. Collects (through surveys and experiments) and organizes data into tables, charts, graphs, and diagrams.

10. Organizes information by using tables, charts, and a variety of graph types with appropriate labels and scales, and interprets displays such as those found in public media.

11. Reads and interprets tables, charts, graphs, and diagrams.

12. Recognizes a wide variety of occupational situations in which information is gathered and displayed, using tables, charts, and graphs.

13. Determines the mean, median, mode, and range of data, and uses these measures to describe the set of data.

14. Applies simple statistical techniques to problem-solving situations.

Strand 3: Measurement and Geometry (32–34% of the test), Chapters 5, 6, and 7

15. Estimates measures in both customary and metric systems.

16. Estimates and solves problems involving measurement, including selecting appropriate tools such as calculators or mental calculation.

17. Applies customary or metric units of measure to determine length, area, volume/capacity, weight/mass, time, and temperature (includes evaluating reasonableness and precision of results, and reading different scales).

18. Identifies items from real life that commonly are measured in metric, in customary, or in both systems of units, as well as recognizing the appropriate-sized units to use.

19. Identifies and differentiates between similar and congruent figures, and identifies figures that have been transformed by rotation, reflection, and translation.

20. Uses proportions to find missing lengths of sides of similar figures, and to enlarge or reduce figures.

21. Solves problems involving similar figures and scale drawings.

22. Graphs points in the coordinate plane, identifies the coordinates, and uses the concept of coordinates in problem situations, such as reading maps.

23. Finds the perimeter and area of plane figures (such as polygons, circles, composite figures) and surface area, and the volume of simple solids (such as rectangular prisms, pyramids, cylinders, cones, spheres).

24. Calculates perimeter and area of plane figures; finds appropriate measures of objects and their models prior to such calculations for basic polygons and circles.

25. Identifies lines, angles, circles, polygons, cylinders, cones, rectangular solids, and spheres in everyday objects.

26. Applies geometric properties—such as the sum of the angles of a polygon property, percent of area of a circle determined by the central angle measure in a pie chart, or parallel sides and angle relations for parallelograms—to practical drawings.

27. Draws and measures angles; determines the number of degrees in the interior angles of geometric figures, such as right and straight angles, circles, triangles, and quadrilaterals; and classifies angles (right, acute, obtuse, complementary, supplementary) and triangles (right, acute, obtuse, scalene, isosceles, and equilateral).

28. Uses the Pythagorean theorem to solve problems (includes selecting appropriate tools such as the calculator).

29. Applies ratios to similar geometric figures, as in scale drawings, as well as with mixtures and compound applications.

Strand 4: Algebra (28–30% of the test), Chapters 8 and 9

30. Simplifies expressions with and without grouping symbols.

31. Evaluates simple algebraic expressions.

32. Substitutes known values in formulas and solves problems with formulas.

33. Identifies and applies mathematics to practical problems requiring direct and inverse proportions.

34. Translates words into simple algebraic expressions and equations.

35. Solves simple equations, including addition, subtraction, multiplication, division, proportions, and two-step equations.

36. Identifies ratio and proportion as they appear in applied situations and solves proportions for missing numbers in applied problems.

37. Solves linear inequalities in one variable and graphs the solution set on the number line.

38. Graphs a linear equation in two variables.

39. Finds the slope and intercepts of a graphed line.

40. Solves problems that involve systems of two linear equations in two variables.

Test-Taking Strategies

What to Do Before the Test

- **Pay attention in class.**

- **Carefully work through the review sections of this book.** Mark any topics that you find difficult, so that you can focus on them while studying and get extra help if necessary.

- **Take the practice tests and become familiar with the format of the GHSGT Mathematics.** When you are practicing, simulate the conditions under which you will be taking the actual test. Stay calm and pace yourself. After simulating the test only a couple of times, you will feel more confident, and this will boost your chances of doing well.

- **Students who have difficulty concentrating or taking tests in general may have severe test anxiety.** Tell your parents, a teacher, a counselor, the school nurse, or a school psychologist well in advance of the test. They may be able to help you learn some useful strategies that will help you feel more relaxed, so that you can do your best on the test.

What to Do During the Test

- **Read all of the possible answers.** Just because you think you have found the correct response, do not automatically assume that it is the best answer. Read through each answer choice to be sure that you are not making a mistake by jumping to conclusions.

- **Use the process of elimination.** Go through each answer to a question and eliminate as many of the answer choices as possible. By eliminating two answer choices, you have given yourself a better chance of getting the item correct, since there will only be two choices left from which to make your guess. Sometimes a question will have one or two answer choices that are a little odd. These answers will be obviously wrong for one of several reasons: they may be impossible given the conditions of the problem, they may violate mathematical rules or principles, or they may be illogical.

- **Work on the easier questions first.** If you find yourself working too long on one question, make a mark next to it on your test booklet and continue. After you have answered all of the questions that you know, go back to the ones you have skipped.

- **Be sure that the answer oval you are marking corresponds to the number of the question in the test booklet.** Since the multiple-choice sections are graded by machine, marking one wrong answer can throw off your answer key and your score. Be extremely careful.

- **Work from answer choices.** You can use a multiple-choice format to your advantage by working backward from the answer choices to solve a problem. This strategy can be helpful if you can just plug the answers into a given formula or equation. You may be able to make an educated guess after eliminating choices that you know do not fit into the problem.

- **If you cannot determine what the correct answer is, guess anyway.** The GHSGT does not subtract points for wrong answers, so be sure to fill in an answer for every question. It works to your advantage because you could guess correctly and increase your score.

The Day of the Test

On the day of the test, you should wake up early (it is hoped after a decent night's rest) and have a good breakfast. Make sure to dress comfortably, so that you are not distracted by being too hot or too cold while taking the test. Make sure to give yourself enough time to arrive at your school early. This will allow you to collect your thoughts and relax before the test, and will also spare you the anguish that comes with being late.

Chapter 1
Numbers and Computation, Part 1

Standards

- Expresses numbers in equivalent and approximate forms and orders these forms by using appropriate tools such as calculators. Number forms include fractions, decimals, percents, second and third powers of whole numbers, scientific notation, and square and cube roots.

- Recognizes, describes, and applies certain patterns for addition and multiplication.

- Selects and uses problem-solving strategies and computational tools (mental computation, calculator, estimation, paper and pencil) to solve simple problems involving career, consumer, and leisure applications, and evaluates reasonableness of results.

About 17 to 19 percent of the 60 questions on the Georgia High School Graduation Test (GHSGT) will be about numbers and computation. In this chapter, you'll learn about equivalent numbers, patterns, and computational tools.

Equivalent number questions will ask you to choose a number with the same value as another number. The two numbers usually will be in different forms. You may be asked, for example, to choose a fraction with the same value as a decimal. Some questions will ask you to determine which of a group of numbers is the greatest. The numbers in these questions often are raised to a power. Other questions will ask you to convert numbers into different forms. For example, you might be asked to convert a percent into a decimal. In this chapter, you'll learn about different forms of numbers. You can use a calculator to answer questions on the GHSGT; therefore, you'll also learn how to use a calculator to determine number forms in this chapter.

Some questions on the GHSGT will ask you about number patterns. Many of these questions refer to the associative and commutative properties of addition and multiplication. You'll learn about these properties in this chapter, too.

Equivalent Numbers

Numbers that are equivalent have the same value. With some numbers, it is easy to see that they are equivalent. For example, you know that $7 = 7$, and you probably know that $\frac{5}{5}$ is equivalent to 1.

Determining whether numbers are equivalent when they are in different forms is more difficult, however. You might not know right away that 8^3 is equivalent to 512.

The best way to determine whether numbers are equivalent is to put them in the same form. The following guidelines will help you do this.

Fractions

A **fraction** represents the number of parts of something that is divided into an equal number of parts. The numerator (top number) of a fraction tells how many parts you have. The denominator (bottom number) tells into how many parts the object is divided. For example, the fraction $\frac{2}{3}$ tells you that you have 2 out of 3 parts.

If the denominators of two fractions are the same, the fraction with the *larger* numerator is the larger fraction. For example, $\frac{3}{7}$ is larger than $\frac{2}{7}$.

Equivalent fractions are fractions that have the same value.

If two fractions have different numerators and denominators, you can determine whether they are equivalent by making the denominators the same. To do this, you must realize that 1 times any number equals the same number (this is discussed later in this chapter). Multiply one of the fractions by the equivalent of 1 so that the

Chapter 1: Numbers and Computation, Part 1   13

denominators of the two fractions are the same. Then compare the results to see whether the fractions are equivalent.

To multiply two fractions, multiply the two numerators to get the new numerator, and then multiply the two denominators to get the new denominator, so $\frac{2}{5} \times \frac{3}{4} = \frac{6}{20}$.

For example, to determine whether $\frac{2}{3}$ and $\frac{4}{6}$ are equivalent, multiply $\frac{2}{3}$ by $\frac{2}{2}$ so it has the same denominator (6) as $\frac{4}{6}$. (Note that $\frac{2}{2}$ is equivalent to 1, so it doesn't change the value of $\frac{2}{3}$.)

$$\frac{2}{3} \times \frac{2}{2} = \frac{4}{6}$$

Therefore, $\frac{2}{3} = \frac{4}{6}$, and the two fractions are equivalent.

Let's try another example: Determine whether $\frac{2}{3}$ is equivalent to $\frac{5}{9}$.

Remember to multiply $\frac{2}{3}$ by $\frac{3}{3}$ so it has the same numerator as $\frac{5}{9}$.

$$\frac{2}{3} \times \frac{3}{3} = \frac{6}{9}$$

$\frac{6}{9}$ is larger than $\frac{5}{9}$, so the fractions are not equivalent.

If you're asked to compare two or more mixed numbers (a mixed number has a whole number and a fraction, such as $1\frac{1}{2}$), the one with the larger whole number is the greater number. For example:

$$2\frac{1}{3} \text{ is greater than } 1\frac{1}{3}$$

If the whole number parts are the same, use the method you just learned to compare the fractional parts to determine which mixed number is larger.

Decimals

A mixed decimal number, such as 3.14, includes a decimal point and has two parts. The part to the left of the decimal point is a whole number, and the part to the right of the decimal point is called a **decimal**. A decimal is not a whole number; it is a portion of a whole number, with a value less than 1. Therefore, the number 3 is greater than the number .33. The decimal .33 can also be written as 0.33, indicating that there is no whole number part to the decimal.

The decimal system is based on the number 10 (this probably has to do with the fact that most humans have 10 fingers). Each digit in a decimal number has a value assigned to its "place." To the left of the decimal point, the digits appear as you are used to seeing them (ones, tens, hundreds, etc.), but to the right of the decimal point they are fractions, so they are tenths, hundredths, thousandths, and so on. (The decimal system is discussed further in the section on scientific notation later in this chapter.)

So, for the decimal number 4.25, the 4 is a whole number. However, the 2 is tenths and the 5 is hundredths, or you could put these last two together and say 25 hundredths. You would read 4.25 as "four point two five," or "four and twenty-five hundredths."

You will likely be asked to compare decimals on the GHSGT. Do you know which is greater, .334 or .3? To determine which decimal is greater, align the decimal points vertically, like this:

.334
.3

Then fill in the empty place values with zeros so both numbers have the same number of digits before you do the comparison:

.334
.300

Which decimal is greater? If you said .334, you're correct! This method is similar to comparing whole numbers, but you must remember to add the zeros to the ends of the decimal fractions so each decimal has the same number of digits.

If you're asked to compare two mixed decimals, the decimal with the greater whole number is always larger. For example, 2.334 is greater than 1.945. If you're asked to compare two mixed decimals with the same whole number, use the method you just learned to compare the decimals to determine which is greater.

Chapter 1: Numbers and Computation, Part 1

For example, to compare 1.4 and 1.36, align the decimal points of the numbers vertically and fill in zeros if necessary:

$$1.40$$
$$1.36$$

Now you can see that 1.4 is definitely greater.

On the GHSGT, you may be asked to **compare a fraction and a decimal**. The best way to do this is to convert the fraction into a decimal. To make a fraction into a decimal, divide the denominator into the numerator. You can use your calculator to do this. Try converting the following fractions into decimals on your calculator.

$$\frac{3}{4} = .75 \qquad \frac{5}{6} = .8333333\ldots$$

Terminating decimals are decimals that stop. For example, .75 is a terminating decimal. **Repeating decimals** keep on going; for example, .8333333.... is a repeating decimal (the 3 keeps repeating). To indicate that a decimal is a repeating decimal, a line usually is placed over the repeating number(s), like this: $.8\overline{3}$.

Percents

A **percent** has a percent sign (%) and refers to how much of a hundred a number is. For example, 75% means 75 out of 100.

Determining which of two (or more) percents is greater is sometimes easy. For example, 75% is obviously greater than 65%. Most often on the GHSGT, however, you will be asked which number in a different form is equivalent to or greater than a percentage. You might be asked, for example, whether 75% is equivalent to $\frac{3}{4}$. (It is!)

As you learned earlier in this chapter, to find an equivalent number or to compare numbers, convert the numbers to the same form. Usually, it is easiest to convert to decimals to compare numbers of different forms.

To convert a percent into a decimal, move the decimal point to the left two places. (These places represent the two zeros in 100.) Look at the following examples:

$$32\% = .32$$
$$75\% = .75$$
$$210\% = 2.10$$

Fill in with zeros if necessary:

$$5\% = .05$$
$$.3\% = .003$$

Now, suppose you need to convert a decimal to a percent. You would move the decimal point two places to the *right*, and add the percentage sign:

$$.20 = 20\%$$

If you need to convert a percent to a fraction, put the percentage over 100. Then **reduce the fraction**, if possible, by the following method: Think of a number that divides evenly into both the numerator and denominator (called a common factor). Do that division, and the results for each part give you the reduced fraction.

So for 20%, the calculation would be $20\% = \frac{20}{100} = \frac{1}{5}$, by dividing both numerator and denominator by 20.

Let's say that you didn't recognize right off that 20 divides into both 20 and 100 in the previous example. Let's say you thought of 10 instead. Then the calculation would be: $20\% = \frac{20}{100} = \frac{2}{10}$. Perhaps now you see that 2 will divide into both the 2 and 10. The result will be $\frac{2}{10} = \frac{1}{5}$, the same result as when you reduced the fraction by using 20 as the common factor: $20\% = \frac{20}{100} = \frac{2}{10} = \frac{1}{5}$.

Powers

At least a few questions on the GHSGT will ask you to raise a number to a certain **power**. The power (indicated by a raised number, called an exponent) tells you how many times the number appears when it is multiplied by itself. Thus, the first power of any number is equal to itself. For example,

$$8^1 = 8$$

Chapter 1: Numbers and Computation, Part 1   17

When you raise a number to the second power, you square the number. When you square a number, you multiply it by itself, as in this example:

$$8^2 = 8 \times 8$$

When you raise a number to the third power, you cube the number. To cube a number, multiply it by itself and then by itself again:

$$8^3 = 8 \times 8 \times 8$$

You can keep raising numbers to higher and higher powers, as in this example:

$8^9 = 8 \times 8 \times 8 \times 8 \times 8 \times 8 \times 8 \times 8 \times 8$, where the number 8 appears nine times.

On the GHSGT, you can use a calculator to raise a number to a power. To square a number, press AC. Then press the number and the x^2 key.

For numbers raised to a higher power, press AC. Then press the number, and then the x^y key, and then the exponent (y). For example, to determine 5^3, press AC-5-x^y-3.

Use your calculator to raise each of the following numbers to the indicated power. See if you get the answers shown here.

$$15^2 = 225$$
$$16^1 = 16$$
$$8^4 = 4{,}096$$
$$20^2 = 400$$
$$9^5 = 59{,}049$$

Scientific Notation

Scientific notation is a type of shorthand for writing numbers that are very large or very small. These numbers contain many zeros. The following table shows the exponents of 10 used in scientific notation. (Note that for powers of 10, the exponent tells you how many zeros follow the 1.)

Number Value	Power of 10
1	10^0
10	10^1
100	10^2
1,000	10^3
10,000	10^4
100,000	10^5
1,000,000	10^6

But what about a number like 2,500,000? How would you write this number in scientific notation? Move the decimal point to the *left* until there is just one digit to the left of it. Here we would move the decimal point until it is between the 2 and the 5. The number of places you move the decimal point (in this case, 6) is the exponent of 10.

$$2{,}500{,}000 = 2.5 \times 10^6$$

Let's try another number. How would you write 32,000 in scientific notation? Remember to move the decimal point until it is in between the 3 and the 2. The number of places you moved it is the exponent of 10.

$$32{,}000 = 3.2 \times 10^4$$

Now, let's work backward. Write out the number 4.8×10^5.

This time you need to move the decimal point to the *right* five places. Remember to fill in zeros.

$$4.8 \times 10^5 = 480{,}000$$

When you work with large numbers on a calculator, the calculator automatically writes the number in scientific notation if there are too many zeros to display. Scientific notation on a calculator looks different, however, because a calculator omits the "10 ×." For example, 2,400,000 might appear as 2.4E6.

Very small numbers (decimals) can also be represented with scientific notation, but in a slightly different way. A negative exponent is used with the number 10 to indicate a decimal.

The following table shows how scientific notation is used to represent very small numbers. (Note that for negative powers of 10, the exponent tells you what "place" the 1 is in, counting to the right from the decimal point.)

Decimal	Power
0.1	10^{-1}
0.01	10^{-2}
0.001	10^{-3}
0.0001	10^{-4}
0.00001	10^{-5}
0.000001	10^{-6}

How do you think you would write 0.0062 using scientific notation? This time, you would move the decimal point to the *right* until it has only one nonzero digit before it. Here you would move the decimal point until it is between the 6 and the 2. You would need to move it three places, so your exponent would be -3.

$$0.0062 = 6.2 \times 10^{-3}$$

Square Roots

The **square root** of a number is the **inverse operation** (opposite) of squaring the numbers (multiplying the number by itself). For example, $\sqrt{144} = 12$.

Not every number is a perfect square. This means you might not always get a whole number when you find the square root. For example,

$$\sqrt{3} \approx 1.73205081$$

Use your calculator to quickly find the square root of a number on the GHSGT. Press the square root key ($\sqrt{}$), then the number, and then the = key. For example, to find the square root of 169, press the $\sqrt{}$ key, then 1-6-9, and then =.

Let's Review 1: Equivalent Numbers

Complete each of the following questions about equivalent numbers. Use the Tip following each question to help you choose the correct answer. When you finish, check your answers with those at the end of Chapter 1.

1. Which answer shows the number that point A represents on the line graph?

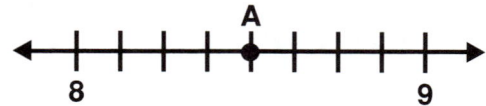

A. $8\frac{1}{2}$

B. $8\frac{5}{8}$

C. $8\frac{3}{4}$

D. $9\frac{1}{4}$

TIP Carefully count the number of places between the number 8 and the number 9. Don't assume there are 10. The number of places should be your denominator.

2. Which is another way to express 144?

A. 12^2

B. 4^4

C. 14.4×10^2

D. 8^3

TIP If you're not sure of the correct answer, begin by eliminating those that you know are incorrect.

3. Which fraction is equal to .45?

A. $\frac{1}{45}$

B. $\frac{9}{20}$

C. $\frac{1}{4}$

D. $\frac{1}{2}$

TIP When you convert a decimal to a fraction, move the decimal point two places to the right and use 100 as the denominator. Then reduce the fraction. You can also solve this problem quickly by using your calculator to find the decimal equivalent of each fraction by dividing the denominators into the numerators and comparing the decimal results.

4. What is .25 expressed as a percent?

A. .25%

B. 2.5%

C. 25%

D. 250%

To convert a decimal to a percent, move the decimal point two places to the right.

Patterns

Numbers have to be added and multiplied in a certain order to solve a problem correctly. On the GHSGT, this is referred to as a **pattern**. Questions about patterns on the GHSGT will often involve the commutative and associative properties. You should also know about identity elements, inequalities, and inverse operations.

Commutative Property

The **commutative property** can be applied to both addition and multiplication. It says that numbers can be added or multiplied in any order.

When the commutative property is applied to addition, for example, $3 + 6$ and $6 + 3$ are the same; they both equal 9.

When you learn about algebra in Chapters 8 and 9 of this book, you will see that letters are sometimes used instead of numbers. For example,

$$x + y = y + x$$

When the commutative property is applied to multiplication, for example, 2×6 and 6×2 are the same; they both equal 12. Using letters, the commutative property can be written two ways:

$$x \times y = y \times x \quad \text{or} \quad xy = yx$$

Questions on the GHSGT might ask you to identify the correct use of the commutative property or they might ask you to identify a situation that demonstrates the commutative property.

A question on the GHSGT about the commutative property might look like the following:

> Daryl plans to spend $6.00 per person for a surprise party for his mother. Ten people are coming to the party. His total cost can be expressed as $6.00 × 10.

Use the commutative property to write an equivalent expression.

A. $6.00 + 10

B. 10 × $6.00

C. $6.00 × .10

D. $\dfrac{\$6.00}{10}$

To correctly answer this question, you need to know that the commutative property of multiplication means that two numbers can be multiplied in any order. The only answer that shows a different order for multiplication of the two numbers is choice B.

Associative Property

The **associative property** can also be applied to both addition and multiplication. It says that the numbers being added or multiplied can be combined in any order.

When it is applied to addition, it means that when you add three numbers, you can add them in any grouping. For example, $5 + 2 + 3$ can be solved like this:

$$(5 + 2) + 3 \text{ or } 5 + (2 + 3)$$

where the parentheses indicate which addition to do first. Both ways add up to 10.

Using letters, we can write the associative property for addition this way:

$$a + b + c = (a + b) + c = a + (b + c)$$

The associative property is true for multiplication also. For example, $2 \times 3 \times 4$ can be solved like this:

$$(2 \times 3) \times 4 \text{ or } 2 \times (3 \times 4)$$

where the parentheses indicate which multiplication to do first. Both ways give the same answer, 24.

Using letters, the associative property for multiplication can be written in two ways:

$$(a \times b) \times c = a \times (b \times c) \text{ or } (ab)c = a(bc)$$

A question about the associative property may be similar to this:

During her first year on the basketball team, Sarah scored x points. During her second year, she scored 65 points, and during her third year, she scored 75 points. Her total points for the three years could be expressed as $x + (65 + 75)$.

Use the associative property to write an equivalent expression.

A. $x + 65 + 75$

B. $65x + 75x$

C. $x(65 + 75)$

D. $(x + 65) + 75$

This question asks you about the associative property of addition. Multiplication should not be involved, so you can eliminate answer choices B and C. Answer choice A does not group two of the three values. Answer choice D is correct.

Identity Elements

Identity elements leave other numbers unchanged when they are combined with them. Think about addition. What can you add to any number and still get that same number? Zero! Zero is the identity element for addition.

The number 1 is the identity element for multiplication. You can multiply any number by 1 and still get that same number. You might see a question on the GHSGT asking you to choose the identity element for either addition or multiplication.

Celia knows that the identity element for addition is 0. What is the identity element for multiplication?

A. 0

B. 1

C. $\dfrac{1}{x}$

D. $\dfrac{0}{x}$

The identity element for multiplication is always 1, so answer choice B is correct.

Other Properties

You might also see questions on the GHSGT about the inverse property and the property of equality or inequality. The **inverse** of something is simply the opposite. The additive inverse of 52 is –52. If you add a number and its inverse together, you get 0, the identity element for addition.

For multiplication, the inverse is a little bit different. If you multiply a number and its inverse, you will get 1, the identity element for multiplication. So, the multiplicative inverse of 7 is $\dfrac{1}{7}$ because if you multiply $\dfrac{7}{1}$ by $\dfrac{1}{7}$, the answer is 1. Note that the multiplicative inverse of a number is also called its **reciprocal**. Zero has no multiplicative inverse, since $\dfrac{1}{0}$ is undefined.

A sample test question may look like this:

Which of the following numbers illustrates the inverse property of addition?

A. $3 + -3 = 1$

B. $3 + -3 = 0$

C. $3 \times -3 = 1$

D. $3 \times -3 = 0$

The correct answer is B because it involves the identity element for addition, 0. Choices C and D can be eliminated right away because they don't involve addition.

When two or more things are equal, as in an equation, an equal sign is used. However, as you know, not all relationships are equal; for example, $7 + 1 \neq 8 + 1$, where $\neq$ means "does not equal." Inequalities usually are indicated by an **inequality** sign. The following are inequality signs you should know:

- $>$ greater than
- $<$ less than
- $\geq$ greater than or equal to
- $\leq$ less than or equal to

Note the following example of a question dealing with inequalities.

Which of the following is a correct statement?

A. $7 > 8$

B. $3 + 4 > 8$

C. $3 + 4 < 5$

D. $3 + 4 < 8$

The correct answer is D: $3 + 4$, or 7, is less than 8. All the other choices are false.

Computational Tools

You might be asked a question on the GHSGT that will ask you to choose the correct computational tool. A **computational tool** is simply a method of solving a problem. The following are computational tools that are commonly used in answer choices on the GHSGT:

- a calculator
- a computer
- mental arithmetic
- paper and pencil

These questions will ask you to choose the computational tool that is most appropriate.

You need to find the exact sum of the prices of eight items in your grocery cart. Which would be the most appropriate method to estimate the exact sum of these items?

A. a computer

B. mental arithmetic

C. paper and pencil

D. a calculator

Imagine a grocery cart with eight items in it, and you need to find the exact sum of their prices. Would you use a computer? You could, but you probably wouldn't have one of these in the grocery store, and it really isn't the best way. Answer choice B is mental arithmetic. It would be difficult to add the prices of eight items by using mental arithmetic. A paper and pencil would work, but it would take a while. A calculator (answer choice D) is the best answer choice, because you could easily add the prices of eight items by using a calculator.

Let's Review 2: Equivalent Expressions

Complete each of the following questions about equivalent expressions. Use the Tip following each question to help you choose the correct answer. When you finish, check your answers with those at the end of the chapter.

1. During one year, a school enrolled a number of new students, expressed as *n*. During the next year, the school enrolled 38 new students. During the following year, it enrolled 92 more students. The principal wrote this expression to show the number of new students enrolled at the school over three years: $n + (38 + 92)$.

Use the associative property to write an equivalent expression.

A. $n \times (38 + 92)$

B. $(n + 38)92$

C. $(n + 38) + 92$

D. $38 + 92n$

TIP Remember that the associative property of addition says that you can add numbers in any order.

2. Which of the following numbers is the additive inverse of −24?

   A. −24

   B. $\dfrac{1}{24}$

   C. 24

   D. 1

Choose the number that is the opposite of −24.

3. Which of the following equations illustrates the commutative property of multiplication?

   A. $x \cdot z = z \cdot x$

   B. $(xy)z = x(yz)$

   C. $\dfrac{xy}{z} = \dfrac{x}{yz}$

   D. $(x + y)z = x(y + z)$

According to the commutative property of multiplication, you can multiply numbers or letters in any order.

4. Which of the following numbers illustrates the inverse property of multiplication?

   A. $-5 + 5 = 0$

   B. $5 \times \dfrac{1}{5} = 0$

   C. $5 \times \dfrac{1}{5} = 1$

   D. $-5 \times 5 = 1$

Remember that for multiplication, a number multiplied by its inverse should equal 1.

5. To calculate the cost of a company's payroll, which method is most appropriate?

A. paper and pencil

B. calculator

C. mental arithmetic

D. computer

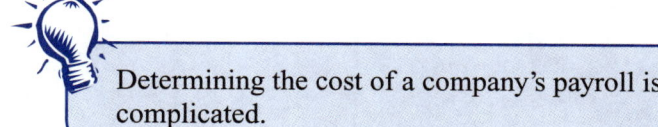
Determining the cost of a company's payroll is complicated.

Chapter 1 Review

Complete each of the following practice problems. Check your answers at the end of this chapter. Be sure to read the answer explanations!

1. Which value is the greatest?

A. 7^3

B. 7^4

C. 8^6

D. 9^5

2. Which answer is .64 written as a percent?

A. .64%

B. 6.4%

C. 64%

D. 640%

3. Jamie knows that the identity element for multiplication is 1. What is the identity element for addition?

   A. 0

   B. 1

   C. $\dfrac{1}{x}$

   D. $\dfrac{0}{x}$

4. What is another way to express 74,000?

   A. 27^2

   B. 34^2

   C. 7.4×10^3

   D. 7.4×10^4

5. Which of the following activities could be used to illustrate the commutative property?

   A. washing a car and then drying the car

   B. standing and then sitting at a desk

   C. putting on a pair of socks and then a pair of shoes

   D. giving a cashier a dollar and then a quarter

6. Which answer shows the number that point B represents on the line graph?

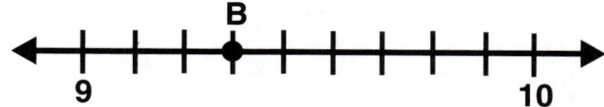

A. $9\dfrac{1}{4}$

B. $9\dfrac{1}{3}$

C. $9\dfrac{3}{10}$

D. $10\dfrac{1}{5}$

7. Which value is the greatest?

A. 3^4

B. 2^5

C. 5^3

D. 6^2

8. Which equation illustrates the associative property of multiplication?

A. $f + g + h = fgh$

B. $(fg)h = f(gh)$

C. $(fg) + h = f + (gh)$

D. $\dfrac{fg}{h} = \dfrac{f}{gh}$

9. Which value is equal to 0.80?

   A. $\dfrac{1}{3}$

   B. $\dfrac{2}{3}$

   C. $\dfrac{4}{5}$

   D. $\dfrac{6}{7}$

10. The equation 2(3x + 4) = (3x + 4)2 is true for all real numbers. Which property does this exemplify?

    A. associative property of addition

    B. association property of multiplication

    C. commutative property of multiplication

    D. commutative property of addition

11. To estimate the sum of the cost of two items, which method is most appropriate?

    A. calculator

    B. computer

    C. mental arithmetic

    D. paper and pencil

TIP: Estimating the cost of two items is not difficult.

Chapter 1: Numbers and Computation, Part 1

Chapter 1 Answers

Let's Review 1: Equivalent Numbers

1. A

To answer this question, count the number of lines between 8 and 9. Remember that your answer should start with 8. Then create a fraction with the number of lines between 8 and 9 as the denominator, and the number of lines to get to point A as the numerator. Reduce your fraction to lowest terms, since $8\frac{4}{8} = 8\frac{1}{2}$.

2. A

If you press the square root key on your calculator and then the numbers 1-4-4, you'll see that the square root of 144 is 12. So, 12^2 is the correct answer. Or you can see that A is the correct answer by pressing the x^2 key and then 1-2, which will give you 144.

3. B

You can answer this question two different ways: you can use your calculator to divide the denominator of each fraction into the numerator, or you can convert .45 into a fraction. To do this, move the decimal point two places to the right and put 45 over 100. Then reduce the fraction (the common factor is 5).

4. C

Remember, to change a decimal into a percent, move the decimal point two places to the right and add the percentage sign.

Let's Review 2: Equivalent Expressions

1. C

The associative property of addition simply means that you can add numbers in any order and get the same result. Answer choice C is the only one that does this.

2. C

For addition, the inverse is the opposite. Remember that when you add a number and its inverse, you should get 0. The additive inverse of –24 is therefore 24, choice C.

3. A

Don't confuse the commutative property of multiplication with the associative property of multiplication. The commutative property of multiplication simply changes the order. It doesn't move parentheses. Note that a dot is the same as a multiplication sign.

4. C

The result of the multiplication of a number by its inverse should be 1.

5. D

Calculating the cost of a company's payroll is difficult. You would need a software program to do this. Therefore, answer choice D, a computer, is the best method.

Chapter 1 Review

1. C

To solve this problem, you can eliminate answer choices A and B, since they are obviously smaller than the numbers in answer choices C and D (the numbers as well as the exponents are smaller). Use your calculator to find the correct answer between choices C and D. $8^6 = 262,144$, whereas $9^5 = 59,049$.

2. C

Remember, to convert a decimal into a percent, move the decimal point two places to the right and add the percentage sign.

3. A

The identity element for addition is 0. Zero added to any number is that same number.

4. D

The correct answer choice to this question uses scientific notation. You need to move the decimal point from the end of the number four places to the left to get between 7 and 4, so the exponent of 10 must be 4.

5. D

Remember that the commutative property means that you can do something in a different order and get exactly the same result. You can't dry a car before you wash it, so answer choice A is not correct. The result isn't the same if you sit first and then stand, so answer choice B is not correct. Putting on a pair of socks always precedes putting on a pair of shoes, so answer choice C is wrong. But if you give a cashier a quarter and then a dollar, or a dollar and then a quarter, the result is the same. Answer choice D is correct.

6. B

Point B is at the third point greater than 9. This gives you the mixed number $9\frac{3}{9}$, or $9\frac{1}{3}$.

7. C

You can use your calculator to figure out the answer to this question quickly. The number 5^3 is 125, and this is the greatest number.

8. B

Remember that with the associative property of multiplication, the parentheses are moved. Therefore, answer choice B is the correct answer. (It is also the only answer that uses multiplication of the three letters.)

9. C

If you convert the decimal 0.80 to a fraction, you get $\frac{80}{100}$. When you reduce this number, you get $\frac{4}{5}$.

10. C

The parentheses have not moved here, just the number 2 (think of the quantity in the parentheses as one "number"). Therefore, it's the commutative property of multiplication.

11. C

It is very easy to estimate the sum of the cost of two items. You could use mental arithmetic to do this.

Chapter 2
Numbers and Computation, Part 2

Standards

- Selects and uses problem-solving strategies and computational tools (mental computation, calculator, estimation, paper and pencil) to solve simple problems involving career, consumer, and leisure applications, and evaluates reasonableness of results.

- Uses estimation strategies such as rounding, front-end estimation, clustering, grouping, adjusting, compensation, and reference point to predict computation results.

- Uses estimation and approximation to check the reasonableness of computational results.

- Determines the amount of money, including price, amounts of change, discounts, sales prices, sales tax, interest, and best buy.

Some Numbers and Computation questions on the Georgia High School Graduation Test will ask you to estimate an answer. You may be asked to choose the best estimate of the sum or difference of a set of numbers, or you may be asked to choose a situation in which an approximate or exact answer is appropriate. Most questions will be about real-life situations, such as having to estimate a discount for an item on sale. You may use a calculator to answer these questions.

Other questions on the GHSGT will be about money. For these questions, you may be asked to figure out how much something costs after a discount, or you may be asked to determine a discount such as 25% off a $100 item. Other questions involving money might ask you to calculate interest and sales tax. You may use a calculator to answer these questions.

Estimation

When you **estimate**, you find the approximate value. You'll learn some of the most common methods of estimation in this chapter.

Rounding

For some estimation questions on the GHSGT, you'll have to round numbers to the same place value. The following numbers are rounded to the tens place:

$$12 + 19$$
$$10 + 20$$

If you add the estimations of these numbers, the answer is 30. The following numbers are rounded to the hundreds place:

$$186 + 342$$
$$200 + 300$$

The estimated answer is 500. If you rounded the same numbers to the tens place, they would look like this:

$$190 + 340$$

Can you estimate this answer in your head? If not, use a calculator. The estimated answer is 530.

These numbers are rounded to the thousands place:

$$1,230 + 4,689$$
$$1,000 + 5,000$$

The estimated answer is 6,000. If you rounded these numbers to the hundreds place, they would look like this:

$$1,200 + 4,700$$

The estimated answer is 5,900.

Chapter 2: Numbers and Computation, Part 2

Round each of the numbers to the tens place to solve this problem:

Estimate the sum of 42, 14, 28, 23, and 29.

If you round the numbers to the tens place, they look like this:

Number	Rounded to the Tens
42	40
14	10
28	30
23	20
29	30

Now add the rounded numbers:

$$40 + 10 + 30 + 20 + 30 = 130$$

The estimated answer is 130.

Clustering

When you are asked to find the sum or difference of a group of numbers, it can help to **cluster** these numbers. This means that you add a few numbers at a time. For example, the numbers in the preceding example could be added two at a time, like this:

$$40 + 10 = 50$$
$$50 + 30 = 80$$
$$80 + 20 = 100$$
$$100 + 30 = 130$$

Front-End Estimation

Another method of estimation is called **front-end estimation**. With this type of estimation, you round and add only the numbers in the leftmost place. Front-end estimation was used to estimate the difference or sum of the following numbers:

$$45,736 - 28,924$$
$$50,000 - 30,000 = 20,000$$

154 + 790
200 + 800 = 1,000

1,241 + 3,880
1,000 + 4,000 = 5,000

Use front-end estimation to solve this problem:

Estimate the difference of 2,945 − 1,523.

If you use front-end estimation, the numbers are rounded as follows:

Number	Rounded Using Front-End Estimation
2,945	3,000
1,523	2,000

Now, use the rounded number to estimate the difference:

3,000 − 2,000 = 1,000

Measurement

Some questions on the GHSGT will ask you to choose the best estimate of the length or size of an object, such as a child's shoe. To answer these questions, you need to picture the size of the **reference point**, or the object being measured. You might be asked to choose whether the object should be measured in inches, feet, yards, or miles. Read this problem:

The best estimate for the height of a doorway is

A. 7 inches.

B. 7 feet.

C. 7 yards.

D. 7 miles.

If you look at or imagine a doorway, you'll see that it is about 7 feet high. It is much taller than 7 inches (answer choice A), and 7 yards (answer choice C) would be too high. And 7 miles (answer choice D) would be much, much too high.

Percentages

You might also be asked to estimate percentages on the GHSGT. You might be asked to estimate a 25% discount on a sweater that costs $31.00. To solve this type of problem, it's usually best to round the amount—in this case the cost of the sweater—to the nearest tens or hundreds, and then solve the problem. Look at the steps in the following problem.

If a sweater, which originally cost $31.00, is selling at a 25% discount, what is the amount of the discount?

1. Round $31.00 to $30.00.

2. Figure out the discount: 10% of $30.00 is $3.00, so 20% of $30.00 is $6.00, and 5% of $30.00 is $1.50.

3. Add $6.00 and $1.50.

4. The approximate discount is $7.50.

Situations

Some test questions will ask you to choose a situation in which an approximate answer is acceptable. Other questions will ask you to choose a situation requiring an exact answer.

Read each of these situations. Think about whether it is more appropriate to estimate or to find the exact answer. Then read the answers that follow.

1. A teacher is planning a party for his class. He needs to determine how many juice boxes, cupcakes, and party favors are needed for the event.

2. Rami plans to buy a new jacket and scarf and checks to see whether he has enough money.

3. Mr. Martinez needs to pay a carpenter to install a patio in his backyard.

4. A coach needs to figure out the number of points a school basketball team scored in a tournament.

Consider these answers:

1. Estimation would be fine for this, since the teacher really doesn't know how much each student will eat or drink.

2. Estimation would also be fine for this, since Rami needs only a general idea of whether or not he has enough money.

3. Mr. Martinez needs to know the exact amount to pay the carpenter.

4. Every point counts in a basketball game, especially in a tournament, so the exact number is needed.

Now let's try this problem:

Choose the one situation in which a result using approximate numbers would be expected.

A. the cost of dinner for two at a restaurant

B. the number of planes flying into an airport during an afternoon

C. the number of parents who attended an open house at a school

D. the number of minutes allowed on a cell phone plan for one month

If you needed to pay a bill at a restaurant, you would need the exact amount, so answer choice A is not correct. The number of planes flying into an airport during an afternoon is very important to know precisely, so estimation would not be best for answer choice B, either. However, an estimate would be fine for the number of parents who attended an open house at a school (answer choice C). The number of minutes allowed on a cell phone plan for one month would also need to be exact, so answer choice D is not correct.

Chapter 2: Numbers and Computation, Part 2

Let's Review 3: Estimation

Complete each of the following questions. Use the Tip following each question to help you choose the correct answer. When you finish, check your answers with those at the end of Chapter 2.

1. **Estimate the sum of 32, 15, 67, 99, and 13.**

   A. 200
   B. 210
   C. 230
   D. 240

   TIP Round 32, 15, 67, and 13 to the nearest ten. Round 99 to 100.

2. **The best estimate for the length of an adult's arm is**

   A. 2 inches.
   B. 2 feet.
   C. 2 yards.
   D. 2 miles.

   TIP Look at your arm. About how long is it?

3. **Sarah earns $7 an hour bagging groceries during the 10 weeks of summer vacation. If she averages 20 hours per week, what is a reasonable estimate of what Sarah will earn during the summer?**

   A. $140
   B. $170
   C. $1,400
   D. $14,000

   TIP Multiply 7 by 20. Then multiply this number by 10.

4. Choose the situation in which a result using approximate numbers would be expected.

A. the minimum height a child must be to ride a ride at an amusement park

B. the final grade-point average for each student in a class

C. the amount of taxes a homeowner must pay each year

D. the number of miles you can ride your bike in two hours

Approximate numbers are appropriate in less formal situations.

5. Estimate the difference: 32,987 – 12,956

A. 10,000

B. 20,000

C. 30,000

D. 40,000

Round both numbers to the nearest ten thousand. Then subtract.

Computing Money

Some questions on the GHSGT will be about money. These questions are about real-life situations. You might be asked to calculate the sale price of a discounted item, sales tax, or the interest on a short-term loan.

Discounts and Sale Prices

Some questions on the GHSGT will ask you to determine the amount of a discount for an item on sale or the sale price of an item. For example, you might be asked to determine a discount of 15% on shoes that cost $45.00. To do this with a paper and pencil, you would multiply 45 by .15 as shown below:

Chapter 2: Numbers and Computation, Part 2

$$
\begin{array}{r}
45 \\
\times\ .15 \\
\hline
225 \\
+\ 450 \\
\hline
6.75
\end{array}
$$

The discount is $6.75.

> On the GHSGT, you can use a calculator to determine a discount. If you wanted to find the amount of a 15% discount on a pair of $45.00 shoes, you would key in 45, then the multiplication sign, and then 15% or .15.

Other questions will ask you to determine the sale price of an item after a discount is applied. Read the following problem:

Megan wants to buy a mirror for her room that usually is priced at $85.00 and is now discounted by 40%. What is the sale price of the mirror?

To solve this problem by using a pencil and paper, you would multiply .40 by 85, as shown here:

$$
\begin{array}{r}
85 \\
\times\ .40 \\
\hline
00 \\
34.00
\end{array}
$$

Remember that $34 is the amount of the discount. This question asks you to find the sale price of the mirror, so you have to subtract 34 from 80:

$$
\begin{array}{r}
85.00 \\
-34.00 \\
\hline
51.00
\end{array}
$$

The sale price of the mirror is $51.00.

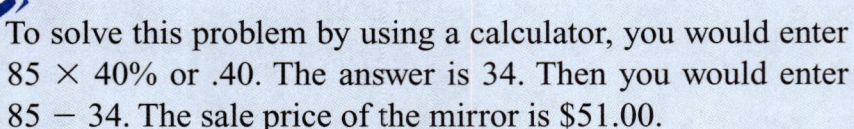

To solve this problem by using a calculator, you would enter 85 × 40% or .40. The answer is 34. Then you would enter 85 − 34. The sale price of the mirror is $51.00.

You determine the sales tax on an item in much the same way that you determine a discount. Read this question:

Gail works in a small hardware store where the cash register does not compute the sales tax. If the sales tax is 7%, what amount should Gail add to a purchase of $10.00?

To answer this question, multiply 10 by .07 or 7%. The amount of sales tax Gail should add to a purchase of $10.00 is $0.70 or 70 cents.

Some questions might ask you to add the sales tax to the cost of item. Read this problem:

Brian wants to buy a bike that costs $125. He knows that he will have to pay 6% sales tax on the bike. How much money, including tax, does Brian need to buy the bike?

To answer this question, you have to calculate the sales tax and add it onto the cost of the bike. Multiply .06 or 6% by 125. When you do this, you get $7.50. Now add this amount to $125, the cost of the bike. The answer is $132.50, which is the amount of money Brian needs to buy the bike.

Interest

When you borrow money, you take a loan. Usually, you're asked to pay interest on the loan. **Interest** is an additional sum of money you must pay in addition to the **principal**, the amount of money you borrowed. Interest is like a fee that you pay to the person or company that lent you the money. On the GHSGT, calculating interest is usually very simple. Read the following problem:

Chapter 2: Numbers and Computation, Part 2

If Alberto borrows $5,000 from a bank at a fixed interest rate of 12% per year, how much interest must he pay if he pays the loan in full at the end of one year?

To solve this problem, you must multiply .12 (that is, 12%) by 5,000. The answer is $600. If Alberto pays the loan in full at the end of one year, he must pay $600 in interest.

Let's Review 4: Computing Money

Complete each of the following questions. Use the Tip following each question to help you choose the correct answer. When you finish, check your answers with those at the end of Chapter 2.

1. **Mario wants to buy a skateboard that is regularly priced at $55 but is now discounted by 15%. What is the sale price of the skateboard?**

   A. $8.25

   B. $46.75

   C. $56.75

   D. $82.50

   TIP Multiply 55 by .15. Then deduct this amount from the price of the skateboard.

2. **Javier works in an ice cream store where the cash register does not compute the sales tax. If the sales tax is 5%, what is the amount Javier should add to a purchase of $11.00?**

   A. $0.45

   B. $0.55

   C. $0.75

   D. $0.85

   TIP Multiply .05 or 5% by 11.

3. If a pair of jeans originally cost $25 and are selling at a 12% discount, what is the amount of this discount?

A. $2.20

B. $3.00

C. $23.00

D. $30.80

TIP Multiply 25 by .12. This is the amount of the discount.

4. Leo's dental plan pays 45% of dental expenses after the deductible of $100 is subtracted. Leo's total dental bill was $380. What is the exact amount the insurance company will pay?

A. $109

B. $126

C. $226

D. $280

TIP Deduct 100 from 380. Then find 45% of this number.

5. If Javier borrows $7,000 to buy a car at a fixed interest rate of 13% per year, how much interest must he pay if he pays the loan in full at the end of two years?

A. $910

B. $920

C. $1,820

D. $1,840

TIP Notice that this question asks you to determine the interest for *two* years. First find the interest for one year, and then multiply this number by two.

Chapter 2 Review

Complete each of the following practice problems. Check your answers at the end of this chapter. Be sure to read the answer explanations!

1. Kate works in a small gift shop where the cash register does not compute the sales tax. If the sales tax is 6%, what is the total amount the customer will pay for a purchase of $25.00?

   A. $1.50

   B. $1.75

   C. $26.50

   D. $41.50

2. The best estimate for the distance from one town to another is

   A. 10 inches.

   B. 10 feet.

   C. 10 yards.

   D. 10 miles.

3. Kristen wants to buy a coat that usually is priced at $125 and now is discounted by 35%. What is the sale price of the coat?

   A. $43.75

   B. $81.25

   C. $90.00

   D. $91.50

4. Miguel earns $15 per lawn to mow grass, and he mows three lawns per week. How much does Miguel earn in four weeks?

   A. $50.00

   B. $90.00

   C. $180.00

   D. $300.00

5. Keisha's eyeglass plan pays 52% of the cost for a pair of glasses after the deductible of $25 is subtracted. Keisha's glasses cost $225. Which is the best estimate of the amount the insurance company will pay?

   A. $90

   B. $100

   C. $120

   D. $140

6. Estimate the sum of 14, 89, 31, 56, and 78.

   A. 190

   B. 270

   C. 350

   D. 400

7. If boots that originally cost $52 are selling at a 25% discount, what is the amount of the discount?

   A. $12

   B. $13

   C. $27

   D. $39

8. If Karen borrows $8,000 from a bank at a fixed interest rate of 14% per year, how much interest must she pay if she pays the loan in full at the end of one year?

   A. $1,120

   B. $2,120

   C. $5,880

   D. $6,880

9. Pedro wants to buy a birthday present for his mother that costs $32.00. He knows he must pay 7% sales tax on the gift. What amount should Pedro add to a purchase of $32.00?

   A. $2.00

   B. $2.24

   C. $22.40

   D. $29.76

10. Choose the situation in which exact numbers would most likely be needed.

    A. The school paper wants a report on the number of students who have a pet.

    B. Michelle is going to a new school. She thinks about the number of letters she will send to her old school friends.

    C. Mrs. Harris, the school principal, has to figure out her annual budget.

    D. A student asks the basketball coach how many lockers are in the locker rooms.

Chapter 2 Answers

Let's Review 3: Estimation

1. C

Round the numbers to the nearest ten: 30, 20, 70, 100, and 10. When you add these numbers, you get 230.

2. B

An adult's arm is generally about two feet or a little longer.

3. C

If you multiply 7 by 20, you get 140. If you multiply 140 by 10 (the number of weeks Sarah worked), you get $1,400.

4. D

It is acceptable to estimate the number of miles you can ride your bike in two hours. The other answer choices are more formal situations, in which an exact answer is required.

5. B

If you round both numbers to the nearest ten thousand, you get 30,000 and 10,000. The difference between these numbers is 20,000.

Let's Review 4: Computing Money

1. B

When you multiply $55 by .15, you get $8.25. When you subtract this amount from the original cost of the skateboard, you get $46.75.

2. B

To find the amount of the sales tax, you need to multiply $11 by .05. The answer is $0.55.

Chapter 2: Numbers and Computation, Part 2

3. B
To solve this problem, you have to multiply $25 by .12. The amount of the discount is $3.00.

4. B
The first step in solving this problem is to subtract 100 from 380. Then multiply .45 by 280. The answer is $126. This is the amount Leo's insurance will pay.

5. C
To solve this problem, you have to multiply $7,000 by .13. The answer is $910. Since you need to find the amount of interest Javier would pay in two years, you need to multiply 910 by 2.

Chapter 2 Review

1. C
To solve this problem you need to multiply $25 by .06. When you do this, you get $1.50. Then add this amount to $25. The answer is $26.50.

2. D
The distance from one town to another—unless they're right next to each other—is probably large, and thus likely to be measured in miles. Ten miles is the best answer.

3. B
To solve this problem, multiply 125 by .35. Then deduct this amount from the original price of the coat. $125 − $43.75 = $81.25

4. C
If you multiply 15 by 3, you get 45. If you multiply 45 by 4, you get 180.

5. B
If you deduct $25 from $225, you get $200. If you multiply $200 by .52, you get $104. The best estimate is B, $100.

6. B

To estimate the sum of these numbers, you need to round them to the nearest ten and add them: $10 + 90 + 30 + 60 + 80 = 270$.

7. B

If you multiply the price of the boots, $52, by .25, the amount of the discount, the answer is $13.

8. A

To solve this problem, you have to multiply $8,000, the amount of money Karen borrowed, by .14, the amount of the interest. The answer is $1,120.

9. B

To find the sales tax on $32, multiply this number by .07, the amount of the sales tax. The answer is $2.24. This is the amount that Pedro should add to the cost of the purchase.

10. C

Answer choice C would require an exact number.

Chapter 3
Data Analysis, Part 1

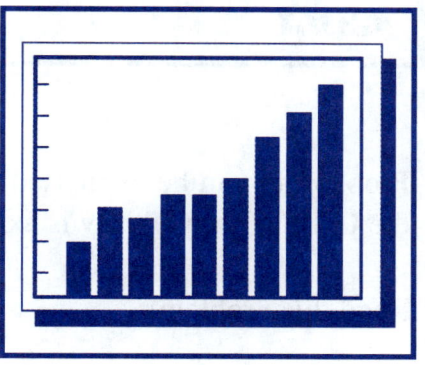

Standards

- Uses probability correctly to predict outcomes of given events, determines the probability of an event through experiments, and differentiates odds from probability.

- Determines the mean, median, mode, and range of data, and uses these measures to describe the set of data.

- Applies simple statistical techniques to problem-solving situations.

In this chapter, you'll learn how to solve data analysis questions involving probability and measures of central tendency.

Probability is the odds (or probability) that an event will happen. Like many of the other questions on the GHSGT, probability problems often will involve real-life situations.

The **measures of central tendency** tested on the GHSGT include mean, mode, median, and range. You'll learn how to determine each of these for a set of data.

Probability

Probability can be determined using this formula:

$$P = \text{number of favorable outcomes}/\text{number of possible outcomes}$$

55

Probability can be expressed as a fraction, a decimal, or a ratio. Most of the time on the GHSGT, probability is expressed as a fraction.

Read this problem:

Find the probability of spinning a 3 on the spinner shown.

A. 0

B. $\dfrac{1}{4}$

C. $\dfrac{1}{2}$

D. 1

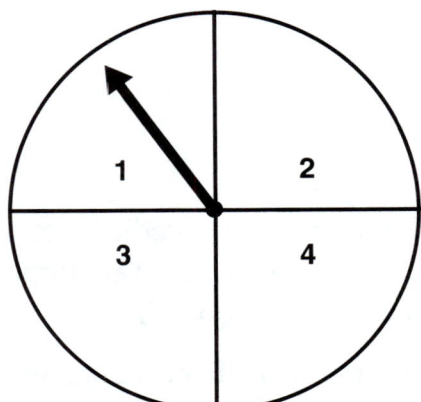

To solve this problem, use the formula shown earlier.

$1 =$ the number of favorable outcomes

$4 =$ the number of possible outcomes

Using this formula, you can see that the probability of spinning a 3 on the spinner is $\dfrac{1}{4}$. Answer choice B is correct. Let's try another problem:

Justine has a bag of 20 marbles. Ten of these marbles are white, three are green, two are blue, and five are yellow. If Justine reaches into the bag and pulls out a marble without looking, what is the probability that she will pull out a yellow marble?

A. 0

B. $\dfrac{1}{20}$

C. $\dfrac{1}{4}$

D. $\dfrac{1}{2}$

Use the probability formula to solve this problem. There are five yellow marbles, so this is the number of favorable outcomes. There are 20 marbles altogether, so this is the number of possible outcomes. The probability that Justine will pull out a yellow marble is $\frac{5}{20}$. This fraction can be reduced to $\frac{1}{4}$. Answer choice C is correct.

On the GHSGT you might be asked to solve a probability problem involving a tree diagram. Read this problem:

Use the tree diagram to predict the probability of flipping two coins and getting all heads or all tails.

A. $\frac{1}{4}$

B. $\frac{1}{2}$

C. 1

D. 2

```
       H    HH
   H <
       T    HT

       H    TH
   T <
       T    TT
```

Look carefully at the tree diagram. The letters to the right of the tree diagram show the possible outcomes. The letter H represents heads, and the letter T represents tails. If you look at these letters, you'll see that there are four possible outcomes. Two of these outcomes are HH and TT, meaning that the coins came up both heads and both tails. If you use the probability formula, you can see that the probability is $\frac{2}{4}$ or $\frac{1}{2}$. Answer choice B is correct. Let's try one more problem:

There are 10 straws in a box; some are white and some are red. The probability of reaching into the box and selecting a white straw is $\frac{2}{5}$. What is the probability of selecting a red straw?

A. $\frac{1}{10}$

B. $\frac{3}{5}$

C. $\frac{4}{5}$

D. 1

You have to work backward to solve this problem. You know there are 10 straws in the box altogether, so the denominator must be 10 before it is reduced. You also know that the probability of reaching into the box and pulling out a white straw is $\frac{2}{5}$. Set up the fractions as shown here:

$$\frac{2}{5} = \frac{x}{10}$$

Ten divided by 5 is 2, so you need to multiply the numerator in the first fraction, 2, by 2. The answer is $\frac{4}{10}$. If 4 out of 10 of the straws are white, then 6 out of 10 of the straws are red. The fraction $\frac{6}{10}$ reduced is $\frac{3}{5}$. The probability of reaching into the box and pulling out a red straw is $\frac{3}{5}$, so answer choice B is correct.

Let's Review 5: Probability

Complete each of the following questions. Use the Tip following each question to help you choose the correct answer. When you finish, check your answers with those at the end of Chapter 3.

1. Use the diagram to predict the probability that a family with three children has three girls or three boys.

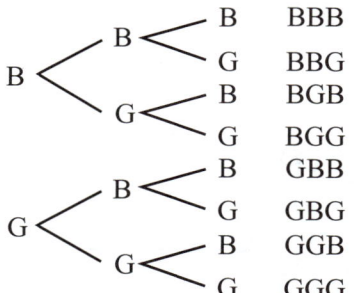

```
              B ─── B   BBB
          B <
              G ─── G   BBG
      B <
              B ─── B   BGB
          G <
              G ─── G   BGG
              B ─── B   GBB
          B <
              G ─── G   GBG
      G <
              B ─── B   GGB
          G <
              G ─── G   GGG
```

A. $\dfrac{1}{4}$

B. $\dfrac{1}{3}$

C. $\dfrac{1}{2}$

D. 1

TIP
Count the number of combinations, using the letters to the right of the tree diagram. Notice that out of all the combinations, two are GGG and BBB, meaning all girls and all boys.

2. Find the probability of spinning "green" on the spinner shown below.

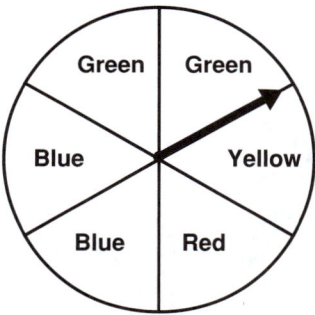

A. 0

B. $\dfrac{1}{4}$

C. $\dfrac{1}{3}$

D. $\dfrac{1}{2}$

TIP
Remember to use the formula for probability and then reduce the fraction. There are six sections on the spinner, and two of these sections are green.

3. A bag contains eight white chips, five red chips, three black chips, two blue chips, and two green chips. If you reach into the jar without looking, what is the probability that you will pull out a red chip?

A. $\dfrac{1}{4}$

B. $\dfrac{1}{3}$

C. $\dfrac{2}{3}$

D. 5

TIP There are 20 chips altogether, and five of these chips are red.

4. Peter is going to roll a six-sided number cube. What is the probability of rolling an even number?

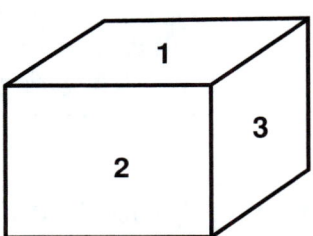

A. $\dfrac{1}{6}$

B. $\dfrac{1}{4}$

C. $\dfrac{1}{3}$

D. $\dfrac{1}{2}$

TIP A six-sided number cube has sides numbered 1, 2, 3, 4, 5, and 6.

5. There are eight jelly beans in a jar; some are pink and some are yellow. The probability of randomly reaching into the box and selecting a pink jelly bean is $\frac{1}{4}$. What is the probability of selecting a yellow jelly bean?

A. $\frac{1}{4}$

B. $\frac{1}{2}$

C. $\frac{3}{4}$

D. $\frac{3}{5}$

Remember that the fraction is reduced. To find out how many yellow jelly beans are in the jar, you need a denominator of 8.

6. If a penny is tossed 10 times and on the first five tosses it comes up heads, what is the probability of getting heads on the sixth toss?

A. $\frac{1}{4}$

B. $\frac{1}{3}$

C. $\frac{1}{2}$

D. 1

When you toss a coin, the odds of it coming up either heads or tails are always the same. It doesn't matter how many times you have tossed the coin.

Mean, Median, Mode, and Range

Some questions on the GHSGT will ask you to analyze data to find measures of central tendency, including the mean, median, mode, and range. **Mean** is another word for "average." To find the mean of a set of numbers, add all the numbers together and divide by the number of items that make up that total. Look at this set of numbers:

$$2, 4, 6, 8, 10$$

To find the mean, you first add all the numbers:

$$2 + 4 + 6 + 8 + 10 = 30$$

Then you divide 30 by the number of items, in this case 5. The mean of these numbers is 6.

The **median** of a set of numbers is the middle. It's not the average, but simply the number in the middle. Look at this set of numbers:

$$10, 4, 2, 8, 6$$

To find the median, you need to put the numbers in order from least to greatest:

$$2, 4, 6, 8, 10$$

When the numbers are in order from least to greatest, you can see that the number 6 is the median. When you have an even number of scores, list from smallest to largest, add the two middle scores together and then divide by two. That will be the median score.

The **mode** of a set of data is the most frequently occurring number. Look at the following numbers:

$$88, 90, 76, 42, 88, 95, 100, 110, 115$$

The mode of these numbers is 88, the only number that occurs more than once.

The **range** of a set of data is the difference between the smallest number and the largest number. Consider these numbers again:

$$88, 90, 76, 42, 88, 95, 100, 110, 115$$

The smallest number is 42 and the largest is 115. To find the range, subtract 42 from 115:

$$115 - 42 = 73$$

The range of this set of numbers is 73.

Now read the problem below, the kind of problem about central tendency that you might see on the GHSGT:

The scores on Mr. Seymour's English test were 98, 60, 88, 87, 96, 79, 80, 58, 76, 99, 80, 58, 76, 99, 90, 87, 62, 76, 89, and 97. What is the range of the scores?

A. 41

B. 58

C. 60

D. 80

To determine the range of this problem, subtract the lowest test score, 58, from the highest, 99.

$$99 - 58 = 41$$

Answer choice A, 41, is the correct answer. Now let's try this problem:

If the mean number of people who attended an amusement park over five days is 25,000, what is the total attendance during the five days?

A. 5,000

B. 50,000

C. 125,000

D. 250,000

To solve this problem, you need to multiply the number of days by the mean. In this case, you would multiply 5 by 25,000. The answer is C, 125,000.

Let's Review 6: Mean, Median, Mode, and Range

Complete each of the following questions. Use the Tip following each question to help you choose the correct answer. When you finish, check your answers with those at the end of Chapter 3.

1. The youngest person in an audience of 300 people is 15 years old. The range of ages is 52 years. What is the age of the oldest member of the audience?

   A. 52
   B. 58
   C. 67
   D. 72

   To answer this question, simply add the range to the age of the youngest person.

2. Jackie spent a total of $100 for six pairs of pants. Later she bought another pair. She spent an average of $19.00 per pair of pants for the seven pairs. What did Jackie pay for the seventh pair?

   A. $13.30
   B. $14.00
   C. $19.00
   D. $33.00

   To solve this problem, multiply $19.00 by 7. Then subtract $100.

3. The total points scored for the Warriors basketball team for each game during the season were 42, 20, 13, 64, 27, 35, 45, 40, 23, 12, 12, and 39. What is their average score?

   A. 12
   B. 31
   C. 35
   D. 52

   You need to find the mean of this data. Add the numbers, and then divide by the total number of scores.

4. Tickets sold at Central Elementary are shown in the chart.

Grade	Number of Tickets Sold
1	246
2	112
3	493
4	98
5	209
6	112
7	190

According to the data in the chart, what is the median of the number of tickets sold per grade level?

A. 112

B. 190

C. 209

D. 395

TIP
Put the number of tickets sold in order from least to greatest. The median is the number in the middle.

5. Renee's World Cultures grades were 84, 85, 95, 88, 92, 100, 82, and 78. What is her average grade?

A. 22

B. 85

C. 88

D. 90

TIP
Add together all her grades, and then divide by 8.

Chapter 3 Review

Complete each of the following practice problems. Check your answers at the end of this chapter. Be sure to read the answer explanations!

1. There are 12 coins in a box; some are nickels and some are pennies. The probability of randomly reaching into the box and pulling out a nickel is $\frac{2}{3}$. What is the probability of pulling out a penny?

   A. $\frac{1}{8}$

   B. $\frac{1}{4}$

   C. $\frac{1}{3}$

   D. $\frac{2}{3}$

2. The oldest person in an audience of 100 is 59. If the range of ages is 43, what is the age of the youngest member of the audience?

   A. 12

   B. 16

   C. 41

   D. 42

3. The weekly salaries of seven employees are $160, $240, $260, $85, $200, $180, and $120. What is the median salary?

   A. $120

   B. $160

   C. $180

   D. $200

Chapter 3: Data Analysis, Part 1

4. Find the probability of spinning a 4 on the spinner.

A. $\dfrac{1}{8}$

B. $\dfrac{1}{4}$

C. $\dfrac{1}{3}$

D. $\dfrac{1}{2}$

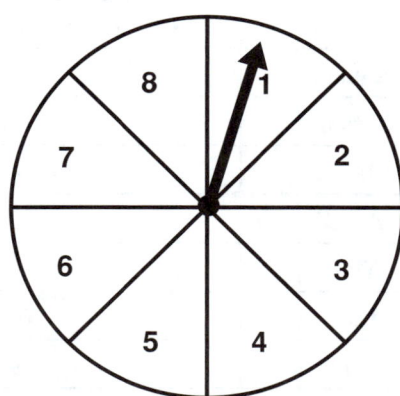

5. Use the tree diagram to predict the probability of flipping a coin and getting heads and of rolling a number cube and getting an odd number.

A. $\dfrac{1}{12}$

B. $\dfrac{1}{4}$

C. $\dfrac{1}{3}$

D. $\dfrac{1}{2}$

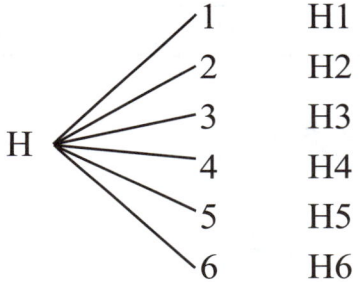

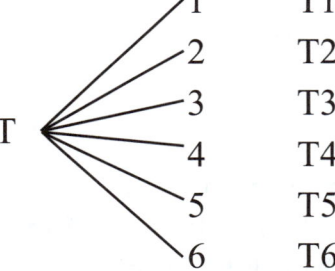

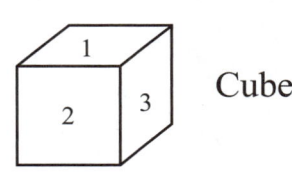

Cube

Coin

6. The average annual temperature in Atlanta is shown in the chart.

Month	Average Annual Temperature
January	41°
February	45°
March	54°
April	62°
May	69°
June	76°
July	79°
August	78°
September	73°
October	62°
November	53°
December	45°

Use the chart to determine the range.

A. 28°

B. 38°

C. 45°

D. 61°

7. Christine has a bag of 25 marbles. Five of these marbles are green, four are blue, three are white, eight are black, and five are yellow. If Christine reaches into the bag, what is the probability that she will randomly pull out a yellow marble?

A. $\frac{1}{8}$

B. $\frac{1}{5}$

C. $\frac{1}{4}$

D. $\frac{1}{3}$

Chapter 3: Data Analysis, Part 1   69

8. Ramon spent a total of $50 for four books. Later he bought another book. He spent an average of $12.25 for each book. What did Ramon pay for the fifth book?

   A. $10.25

   B. $11.25

   C. $12.25

   D. $13.25

9. Kayla recorded the hours she spent studying each week for five weeks. She listed these hours in the chart.

Week	Hours Spent Studying
1	18
2	12
3	15
4	5
5	10

   What is the average number of hours Kayla spent studying?

   A. 10

   B. 12

   C. 13

   D. 15

10. The miles Michelle ran in a week are 2, 3, 3, 4, 4, 3, and 6. What is the mode number of miles Michelle ran?

   A. 3

   B. 4

   C. 5

   D. 6

Chapter 3 Answers

Let's Review 5: Probability

1. A

If you look at the letters to the right of the diagram, you can see that there are eight possibilities, or combinations of children. Out of these eight, there is one possibility that the three children will be all girls, and one possibility that the three children will be all boys. So, the probability that the children will be either all girls or all boys is $\frac{2}{8}$. When you reduce this fraction, the answer is $\frac{1}{4}$.

2. C

The spinner is divided into six sections, and two of these sections are green. So the probability that the spinner will land on green is $\frac{2}{6}$ or $\frac{1}{3}$.

3. A

The bag contains 20 chips altogether, and five of these chips are red. The probability of pulling out a red chip is $\frac{5}{20}$ or $\frac{1}{4}$.

4. D

If the number cube has six sides and is numbered 1, 2, 3, 4, 5, and 6, three of the six sides are even. Therefore, the probability of rolling an even number is $\frac{1}{2}$.

5. C

The question tells you that there are eight jelly beans in a jar, so the denominator should be 8. The question also says that the probability of selecting a pink jelly bean is $\frac{1}{4}$ or $\frac{2}{8}$. If you subtract 2 from 8, the total number of jelly beans, you get 6. So the probability of choosing a yellow jelly bean is $\frac{6}{8}$ or $\frac{3}{4}$.

6. C

If you toss a penny, the odds that it will come up heads or tails are always $\frac{1}{2}$, regardless of how many times you toss the penny.

Let's Review 6: Mean, Median, Mode, and Range

1. C

To solve this problem, add the range, 52, to the age of the youngest person, 15. The answer is 67.

2. D

To answer this question, you need to multiply $19 by 7, which is 133. When you subtract 100 from 133, you get 33, the correct answer.

3. B

When you add all the numbers, you get 372. When you divide 372 by the number of scores, 12, you get 31, the correct answer.

4. B

If you put the numbers in order from least to greatest, you'll see that 190 is in the middle. This number is the median.

5. C

If you add together all of Renee's grades, you get 704. When you divide this number by 8, you get 88.

Chapter 3 Review

1. C

The question tells you that there are 12 coins in a box and some are nickels and some are pennies, and the probability of randomly reaching into the box and pulling out a nickel is $\frac{2}{3}$. You know that the denominator must be 12, so the probability of pulling out a nickel before you reduce the fraction is $\frac{8}{12}$. Therefore, the probability of pulling out a penny is $\frac{4}{12}$ or $\frac{1}{3}$.

2. B

To answer this question, you subtract the range from the age of the oldest person.

3. C

If you put the salaries in order from least to greatest, you'll see that $180 is the median salary.

4. A

The spinner has eight sections, and only one section is numbered 4, so the probability of spinning a 4 is $\frac{1}{8}$.

5. B

There are 12 possibilities, and three of them involve rolling heads and an odd number on the number cube. So, the probability that this will happen is $\frac{3}{12}$ or $\frac{1}{4}$.

6. B

To find the range in the average annual temperature, subtract the lowest temperature, 41°, from the highest temperature, 79°.

7. B

There are 25 marbles in the bag and five of them are yellow. Therefore, the probability of choosing a yellow marble is $\frac{5}{25}$ or $\frac{1}{5}$.

8. B

To answer this question, multiply the average, $12.25, by 5, and then subtract 50 from the amount.

9. B

When you add all the hours Kayla spent studying, you get 60. When you divide 60 by 5, you get 12.

10. A

The mode is the most frequently occurring number. In this case, it is the number 3.

Chapter 4
Data Analysis, Part 2

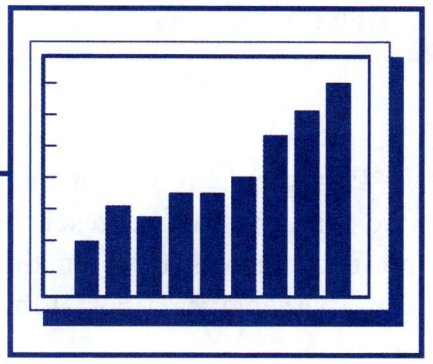

Standards

- Organizes information by using tables, charts, and a variety of graph types with appropriate labels and scales, and interprets such displays as those found in public media.

- Reads and interprets tables, charts, graphs, and diagrams.

- Recognizes a wide variety of occupational situations in which information is gathered and displayed, using tables, charts, and graphs.

In this chapter, you'll learn how to analyze data represented in charts, tables, and graphs. Some questions on the GHSGT will ask you what kind of chart or graph is best to display a certain type of data. These questions are often about pictographs, bar graphs, circle graphs, and line graphs. You'll learn about each of these graphs in this chapter.

Other questions will ask you to correctly interpret data that is represented in graphs. These test questions are often about bar graphs, line graphs, Venn diagrams, and charts. You'll learn how to answer these types of questions in this chapter, too. Since all questions on the GHSGT are multiple choice, you will not be asked to construct a graph.

Pictographs

A **pictograph** uses symbols to display data. A symbol represents a number that is shown in a key that is usually on the side of the graph.

Pictographs are great for illustrating large numbers with large differences, since one symbol could represent many of something. Pictographs don't illustrate trends, however, such as how something changes over a period of time. Pictographs also don't show percentages. Look at this pictograph:

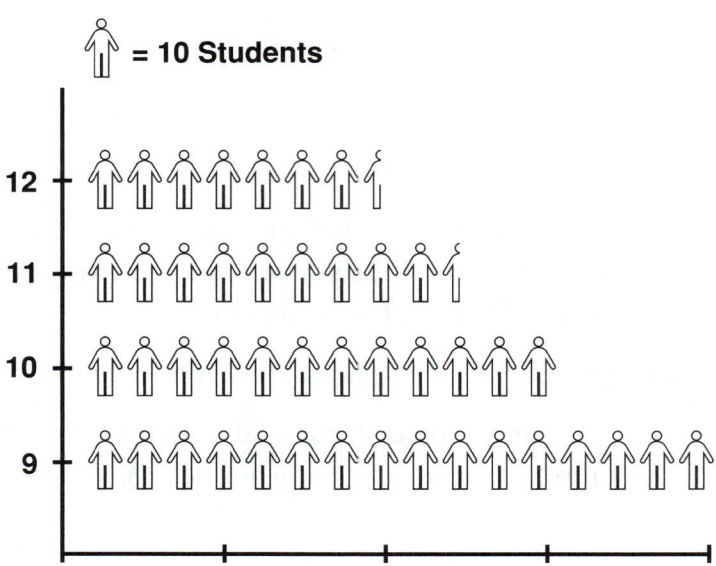

In this pictograph, each picture of a student stands for 10 students. Therefore, there are 160 students in grade 9, 120 students in grade 10, 95 students in grade 11, and 75 students in grade 12. If the high school were very large, each picture of a student might stand for 20 or 50 students. Note that this pictograph doesn't show small differences. The number of students is displayed in fives and tens and not ones.

Bar Graphs

In a **bar graph**, the height or length of a bar shows the number of something. The higher or longer the bar, the greater the value. A bar graph has an x- and y-axis. It is a good way to show comparisons and can show trends, such as changes in sales over time.

Although a bar graph can have either vertical or horizontal bars, most bar graphs on the GHSGT have vertical bars. Look at this bar graph. It shows the number of cars manufactured at a factory over five years.

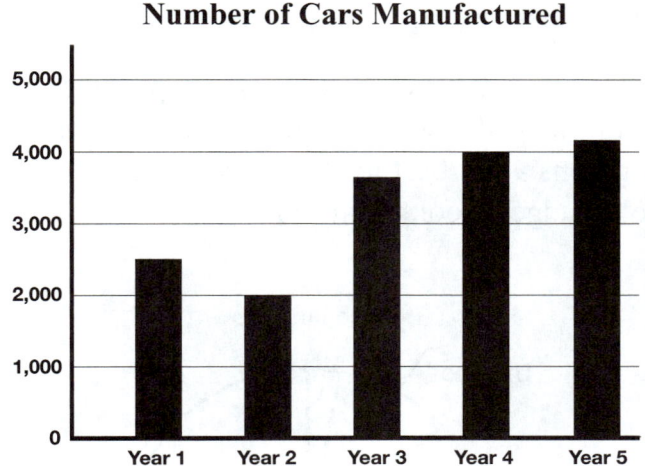

In this bar graph, Years 1 through 5 are listed on the *x*-axis and the number of cars manufactured are listed on the *y*-axis. You can see by just glancing at the graph that the greatest number of cars were manufactured in Year 5 and that except for Year 2, the number of cars manufactured increased each year.

Line Graphs

A **line graph** is a very popular type of graph that compares two variables—one along the *x*-axis and one along the *y*-axis. Unlike in a bar graph, the two variables being compared in a line graph are closely related; a change in one variable is associated with a change in the second variable. A line graph is a great way to show trends. Look at this line graph:

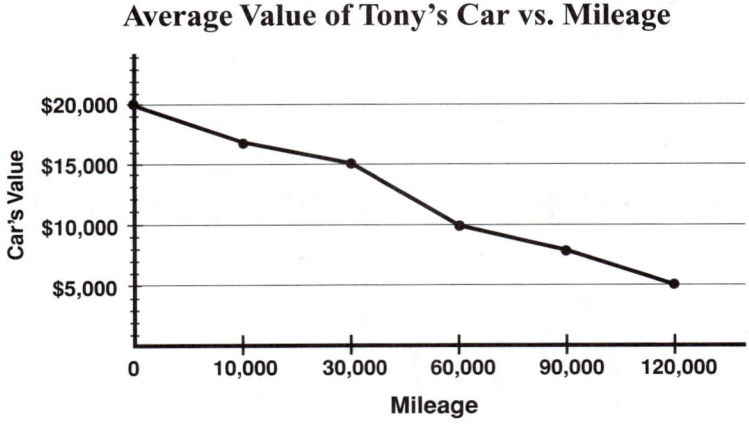

You can see from this line graph that as the number of miles on Tony's truck increases, the value of the truck decreases.

Circle Graphs

A **circle graph**, also called a pie chart or pie graph, is often used to display the division of a whole or parts of a whole. Data in a circle graph is often displayed in percentages. Circle graphs work best to show large divisions, such as the division of money in a household budget. Look at this circle graph:

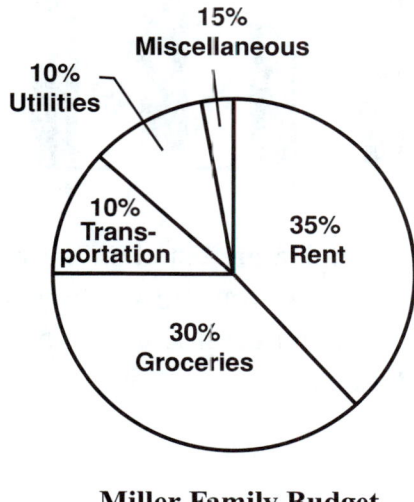

Miller Family Budget

You can see from this circle graph that the Miller family spends most of its monthly income on rent and groceries.

Venn Diagrams

Some questions on the GHSGT may ask you to interpret data displayed in a Venn diagram. A **Venn diagram** is made up of two or more overlapping circles and is used to display relationships between two or more sets of data. Venn diagrams are a great way to show similarities and differences. Look at the Venn diagram shown here. If this diagram were filled in, it would compare two sets of a data, A and B. The part of the circle that overlaps, C, would list ways that the sets of data are alike. Traits unique to each set of data would be in the part of the circles that do not overlap.

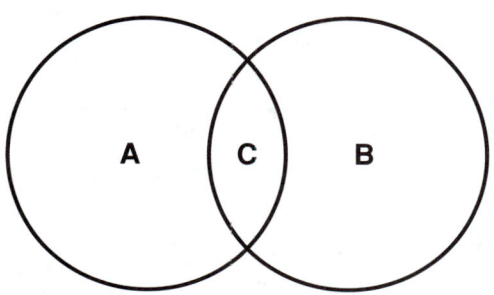

A Venn diagram comparing three sets of data would look like the one shown below. Note that the ways in which A and B are alike would be listed where circle A and circle B overlap (parts D and G). The ways in which A and C are alike would be listed where circles A and C overlap, (parts E and G) and the ways in which B and C are alike would be listed where circles B and C overlap (parts F and G). The ways in which A, B, and C are alike would be listed in the small area where all three circles overlap (part G).

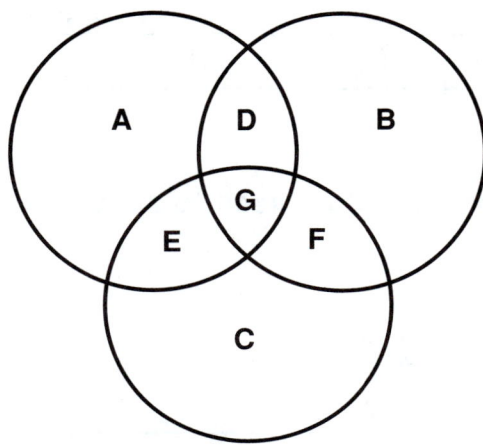

Let's Review 7: Graphs and Venn Diagrams

Complete each of the following questions. Use the Tip following each question to help you choose the correct answer. When you finish, check your answers with those at the end of Chapter 4.

1. **Which kind of graph is best used to show how three sets of a data are alike and different?**

    A. circle graph

    B. bar graph

    C. Venn diagram

    D. line graph

 Only one type of graph or diagram discussed in this chapter shows how things are alike and different.

2. The results of a survey asking how many people lived in a town over five years are shown in the chart.

Year 1	20,000
Year 2	25,000
Year 3	30,000
Year 4	32,000
Year 5	35,000

Which type of graph should be used to show the results of the survey?

A. bar graph

B. circle graph

C. pictograph

D. line graph

Eliminate answer choices that you know are incorrect. Then, if you're still unsure, go back and reread the description of the remaining choices.

3. Which kind of graph is best used to show a percent of share of a total?

A. circle graph

B. line graph

C. pictograph

D. bar graph

If you're not sure of this answer, go back and reread the description of each type of graph.

4.

Extra-Credit Points in English

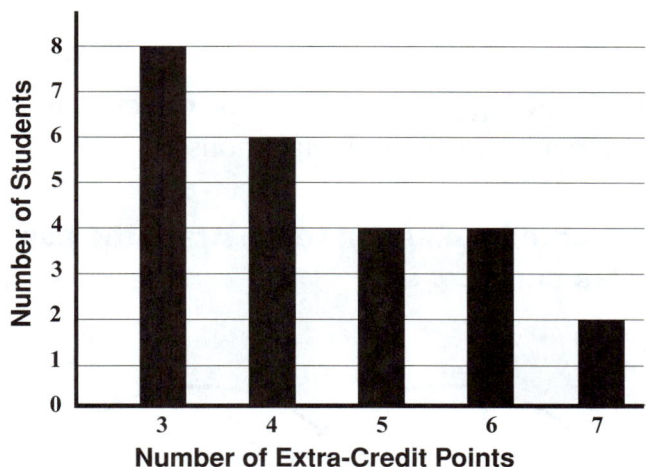

The distribution of extra-credit points in Ms. Washington's English class is shown on this graph. How many more students received five extra-credit points than seven extra-credit points?

A. 2

B. 3

C. 4

D. 5

To answer this question, look at the bar for five extra-credit points and the bar for seven extra-credit points. Subtract the number of students receiving five extra-credit points from the number of students receiving seven.

5. The distribution of extra-credit points in Ms. Washington's English class is illustrated by the graph above. What is the ratio of students receiving three points to students receiving five points?

A. 1:3

B. 2:1

C. 1:2

D. 3:1

A ratio represents a proportion. Find the number of students receiving three points. Then find the number of students receiving five points. What ratio represents this proportion?

Chapter 4 Review

Complete each of the following practice problems. Check your answers at the end of this chapter. Be sure to read the answer explanations!

1. **Mr. Sassy constructed a diagram to illustrate the number of seniors enrolled in honors courses.**

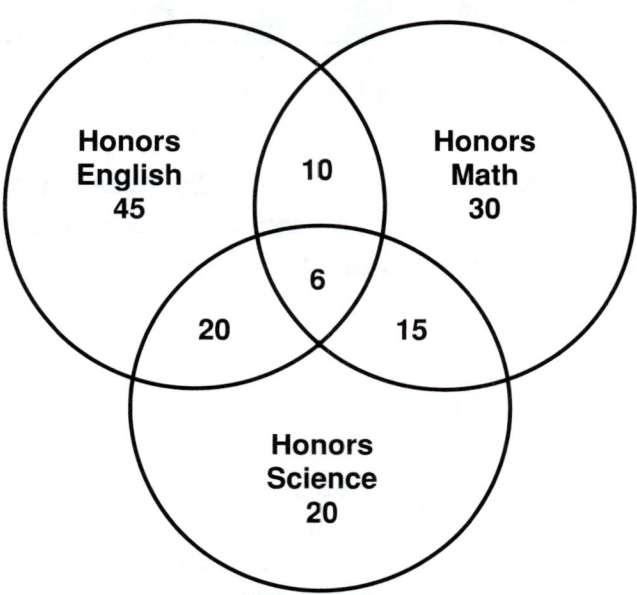

How many seniors are enrolled in Honors Math?

A. 30

B. 40

C. 61

D. 95

2. **How many seniors take all three honors classes: Honors English, Honors Math, and Honors Science?**

A. 6

B. 10

C. 15

D. 20

3. **Which kind of graph is best used to show the relationship between two variables?**

A. bar graph

B. pictograph

C. line graph

D. circle graph

4. **Kevin is making a pictograph to give to his principal showing the number of students in each grade in favor of starting an environmental club. He summarized his data as follows:**

Grade	Number of students in favor of starting an environmental club
Freshmen	50
Sophomores	25
Juniors	100
Seniors	125

In Kevin's pictograph, how many students could best be represented by this symbol?

A. 2

B. 10

C. 25

D. 100

5. **Number of Hours of Community Service**

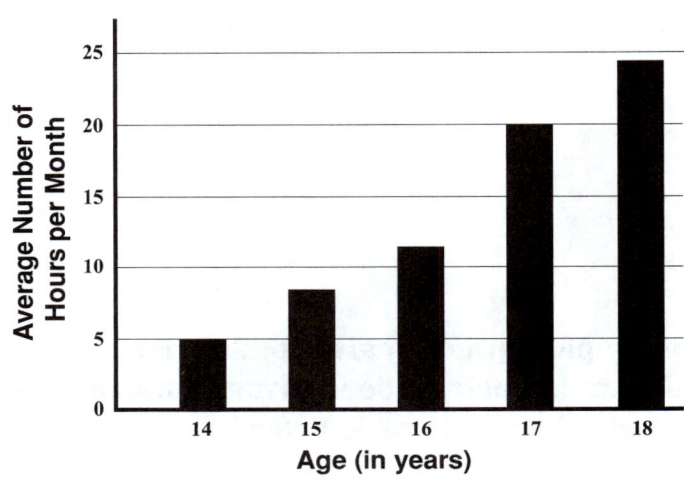

According to the bar graph above, how many hours per month on average did the 17-year-olds work?

A. 8

B. 12

C. 20

D. 24

6. **Refer to the bar graph. How many more hours did the 18-year-olds work than the 14-year-olds?**

   A. 5 hours

   B. 19 hours

   C. 24 hours

   D. 29 hours

7. **The results of a poll asking, "What kind of topping do you like on your pizza?" are shown in the chart.**

Pepperoni	50%
Extra cheese	15%
Mushrooms	5%
Olives	5%
Peppers	15%
Sausage	10%

   Which type of graph should be used to show the results of the poll?

   A. bar graph

   B. circle graph

   C. line graph

   D. pictograph

8. **What is the ratio of those who like olives to those who like sausage?**

   A. 1:2

   B. 2:1

   C. 1:4

   D. 4:1

Chapter 4 Answers

Let's Review 7: Graphs and Venn Diagrams

1. C
A Venn diagram is the best kind of diagram to tell how three things are alike and different.

2. A
A bar graph is the best to show the information given in the question, since it is a good way to show changes over time.

3. A
A circle graph best shows parts of a whole or percents of a total.

4. A
Four students received five extra-credit points, and two students received seven extra-credit points. If you subtract these two numbers, the answer is two.

5. B
Eight students received three extra-credit points, and four students received five extra-credit points, so the ratio is 2:1.

Chapter 4 Review

1. C
To answer this question, you have to look at the number of seniors enrolled in Honors Math, which is $30 + 10 + 6 + 15 = 61$.

2. A
For this question, you need to look at the number in the part of the diagram where all three circles overlap. The answer is six.

3. C

A line graph is the best kind of graph to show the relationship between two variables; one variable will be displayed on the *x*-axis and the other on the *y*-axis.

4. C

If you don't know the answer to this question, you could figure it out by process of elimination. Two (answer choice A) is not the best number to be represented by the symbol, since the number of students in each grade is a multiple of five. The number ten (answer choice B) would work, but there is a better answer choice. Twenty-five (answer choice C) is the best answer, because it can be divided into the number of students in each grade. One hundred (answer choice D) is too large.

5. C

To find the answer to this question, you have to look at the average number of hours the 17-year-olds worked. It's 20.

6. B

The 18-year-olds worked 24 hours and the 14-year-olds worked five hours. If you subtract 5 from 24, you get 19.

7. B

A circle graph is the best choice to show the results of this poll, since it shows parts of a whole.

8. A

Five percent of those polled chose olives as their favorite topping, and 10 percent of those polled chose sausage. The ratio is 1:2.

Chapter 5
Measurement and Geometry, Part 1

Standards

- Estimates measures in both customary and metric systems.
- Identifies items from real life that are commonly used in metric, in customary, or in both systems of units, and recognizes the appropriate-sized units to use.

In this chapter, you will learn how to answer questions on the GHSGT about measurement. Test questions on units of measure might ask you to measure area, capacity, length, mass, temperature, time, volume, and weight. Like most the other questions on the test, these questions will involve real-life situations similar to those you might encounter at home, at work, or at school.

There are two commonly used systems of measurement: customary and metric. You'll learn about both of these in this chapter.

Customary Measures

Customary units refers to U.S. Customary units, the units of measurement most often used in the United States. On the GHSGT you will be asked to choose the best unit of measurement to measure something, such as the length of your arm, a feather, or the distance across town. Study the customary units of measurement in this section for length, weight, capacity, and time.

Measures of Length

1 foot = 12 inches

1 yard = 3 feet

1 yard = 36 inches

1 mile = 5,280 feet

1 mile = 1,760 yards

Measures of Weight (Mass)

1 pound = 16 ounces

1 ton = 2,000 pounds

Measures of Capacity

1 cup = 8 fluid ounces

1 pint = 16 fluid ounces

1 pint = 2 cups

1 quart = 2 pints

1 gallon = 4 quarts

Temperature (Fahrenheit)

32°F   Freezing

212°F   Boiling

Chapter 5: Measurement and Geometry, Part 1

Let's Review 8: Measurement

Complete each of the following questions. Use the Tip following each question to help you choose the correct answer. When you finish, check your answers with those at the end of Chapter 5.

1. **To determine the weight of an automobile, which is the most appropriate unit of measure?**

   A. gallons

   B. ounces

   C. quarts

   D. tons

   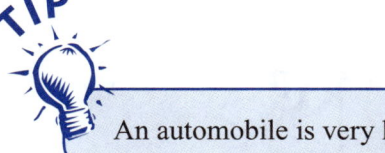
   TIP: An automobile is very heavy.

2. **Jerry wants to find the length of a pencil. Which would be the best unit of measurement?**

   A. inches

   B. feet

   C. yards

   D. miles

   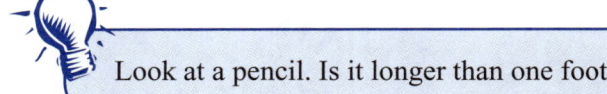
   TIP: Look at a pencil. Is it longer than one foot?

3. **A good estimate for the weight of a three-year-old child would be**

   A. 150 pounds.

   B. 100 pounds.

   C. 30 pounds.

   D. 3 pounds.

   TIP: Think about your own weight. A three-year-old child's weight would be much less than this.

4. If a bag has 10 pounds of flour, how many ounces does it contain?

   A. 16 ounces
   B. 160 ounces
   C. 1,600 ounces
   D. 16,000 ounces

 There are 16 ounces in a pound, so multiply this amount by 10.

5. If a car travels at 50 miles per hour, how many miles will it travel in six hours?

   A. 30 miles
   B. 300 miles
   C. 3,000 miles
   D. 30,000 miles

 To solve this problem, multiply 50 by 6.

6. Given that water boils at 212°F and freezes at 32°F, what would be the most likely temperature for the inside of a refrigerator?

   A. 100°F
   B. 90°F
   C. 45°F
   D. 25°F

 Food in a refrigerator isn't frozen.

Metric Measures

The metric system was developed in France in the late 1700s. Although the customary system is still used in the United States, many other countries and the scientific community throughout the world use the metric system. Some questions on the GHSGT will be about the metric system.

Measures of Length

1 decimeter	=	$\frac{1}{10}$ meter
1 centimeter	=	10 millimeters
1 meter	=	1,000 millimeters
1 meter	=	100 centimeters
1 dekameter	=	10 meters
1 kilometer	=	1,000 meters

Measures of Weight (Mass)

1 gram	=	1,000 milligrams
1 dekagram	=	10 grams
1 hectogram	=	100 grams
1 kilogram	=	1,000 grams
1 metric ton	=	1,000,000 grams

Measures of Capacity

1 deciliter	=	$\frac{1}{10}$ liter
1 liter	=	1,000 milliliters
1 dekaliter	=	10 liter
1 hectoliter	=	100 liters
1 kiloliter	=	1,000 liters

Let's Review 9: Metric Measurement

Complete each of the following questions. Use the Tip following each question to help you choose the correct answer. When you finish, check your answers with those at the end of Chapter 5.

1. Martina wants to find the distance from Burton to Cedartown. Which would be the best unit of measurement?

   A. centimeter

   B. kilometer

   C. meter

   D. millimeter

   Burton and Cedartown are two cities in Georgia. Choose the largest unit of measurement.

2. Megan wants to find the thickness of a piece of paper. Which would be the best unit of measurement to use?

   A. centimeter

   B. kilometer

   C. meter

   D. millimeter

   The thickness of a piece of paper is very small. Choose the smallest unit of measurement.

3. The mass of a carton of milk can best be measured in

   A. grams.

   B. hectograms.

   C. kilograms.

   D. milligrams.

   Eliminate units of measurement that are either very small or very large.

Chapter 5: Measurement and Geometry, Part 1

4. To determine the mass of a refrigerator, which is the most appropriate unit of measure?

   A. grams

   B. centigrams

   C. dekagrams

   D. kilograms

   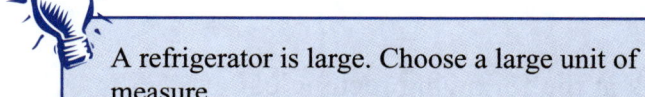 A refrigerator is large. Choose a large unit of measure.

5. A bag has 1,200 grams of grain. How many kilograms does it contain?

   A. between .01 and .02 kilograms

   B. between .1 and .2 kilograms

   C. between 1 and 2 kilograms

   D. between 2 and 3 kilograms

   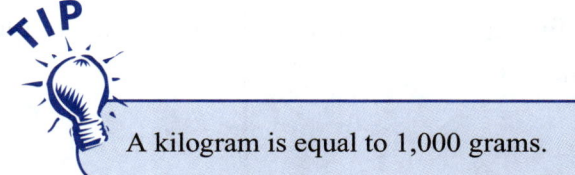 A kilogram is equal to 1,000 grams.

Time

Some test questions will be about time. You will have to determine how much time has passed. For example, you might be asked to solve a problem like this:

Brianna starts work at 11:00 A.M. and stops at 6:30 P.M. If she takes 30 minutes for lunch, what is the length of her workday?

Begin solving this problem by determining how many hours are from 11:00 A.M. until noon—one hour. Then determine how many hours are from noon until 6:30 P.M.—6.5 hours. Then add the first hour to this: 1 hour + 6.5 hours is 7.5 hours. Now subtract 30 minutes, the amount of time Brianna takes for lunch. $7.5 - .5 = 7$.

Brianna's workday is seven hours long.

Area

For some questions on the GHSGT, you will have to determine the area of a figure. Use this formula to find the area:

$$\text{Area} = \text{length} \times \text{width, or } A = lw$$

Look at the following rectangle:

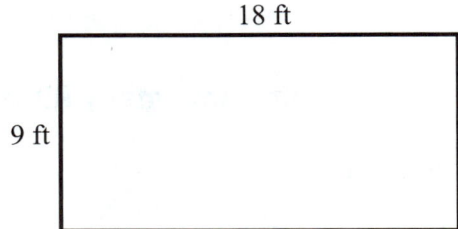

To find the area of this rectangle, plug the length and width into the formula you just learned:

$$A = lw$$
$$A = 18 \text{ ft} \times 9 \text{ ft}$$
$$A = 162 \text{ ft}^2$$

Notice that the area is expressed in square feet (ft^2).

To find the area of a circle, use this formula, where π is pi:

$$\text{Area} = \pi \times \text{radius}^2 \text{ or}$$

$$A = \pi r^2$$

Look at this circle:

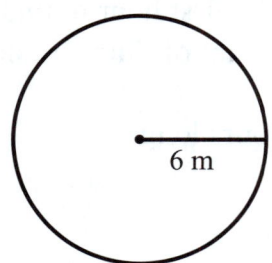

To find the area of this circle, plug the radius into the formula you just learned, which is usually given to you on the test. Use 3.14 for pi (π) and round your answer to the nearest whole number.

$$A \approx (3.14)(6m)^2$$
$$A \approx (3.14)36m^2$$
$$A \approx 113m^2$$

Volume

To find the volume of a figure, use this formula, which is usually provided on the GHSGT:

$$\text{Volume} = \text{length} \times \text{width} \times \text{height}$$

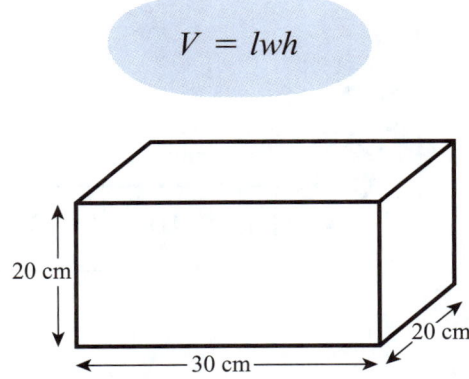

$$V = lwh$$

To find the volume of the rectangular solid shown, substitute the measurements of the length, width, and height into the formula you just learned:

$$V = 30\,\text{cm} \times 20\,\text{cm} \times 20\,\text{cm}$$
$$V = 12{,}000\,\text{cm}^3$$

To find the volume of a cylinder, you would use the formula $V = \pi r^2 h$. To find the volume of a cylinder with a radius of 3 and a height of 5, substitute these values into the formula:

$$V \approx (3.14)3^2 \times 5$$
$$V \approx (3.14) \times 9 \times 5$$
$$V \approx 141$$

Let's Review 10: Time, Area, Volume

Complete each of the following questions. Use the Tip following each question to help you choose the correct answer. When you finish, check your answers with those at the end of Chapter 5.

1. What is the volume of the box pictured here?

   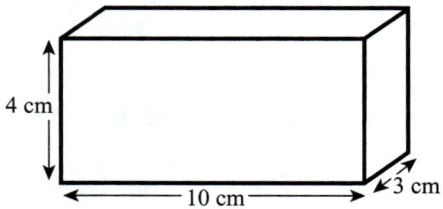

   A. 17 cm³

   B. 40 cm³

   C. 70 cm³

   D. 120 cm³

   TIP: Multiply 10 × 4 × 3.

2. A rotating sprinkler is used to water a yard. The radius of the area being sprayed is 8 feet. What is the approximate wet area of the yard? (use $A = \pi r^2$ and $\pi = 3.14$)

   A. 20 square feet

   B. 25 square feet

   C. 64 square feet

   D. 201 square feet

   TIP: Remember to square the radius.

3. What is the area of the square shown here?

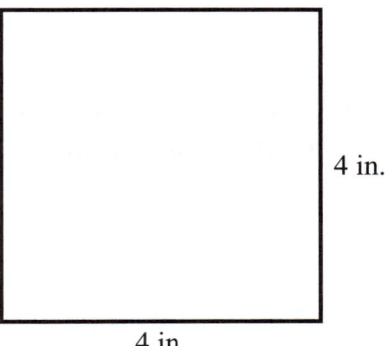

4 in.

4 in.

A. 16 square inches
B. 20 square inches
C. 24 square inches
D. 64 square inches

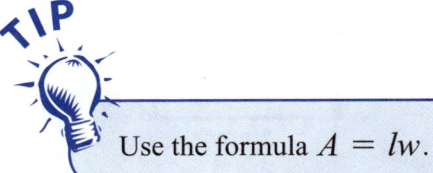

TIP
Use the formula $A = lw$.

4. A seminar begins at 9:00 A.M. and ends at 5:00 P.M. If participants take 45 minutes for lunch, what is the length of the seminar?

A. 7 hours
B. 7.25 hours
C. 7.5 hours
D. 8 hours

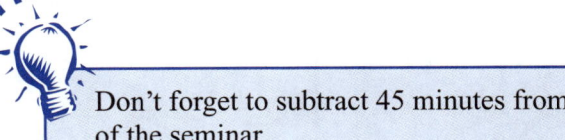

TIP
Don't forget to subtract 45 minutes from the length of the seminar.

Chapter 5 Review

Complete each of the following practice problems. Check your answers at the end of this chapter. Be sure to read the answer explanations!

1. What is the volume of the box pictured here?

   A. 15 cm³

   B. 25 cm³

   C. 125 cm³

   D. 250 cm³

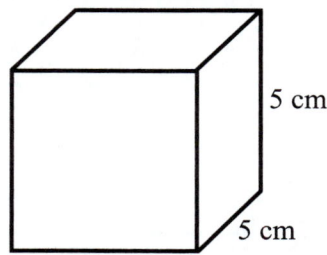

2. Nancy starts her workday at 2:30 P.M. and stops at 11:00 P.M. If she takes 30 minutes for lunch, what is the length of her workday?

   A. 7.5 hours

   B. 8 hours

   C. 8.5 hours

   D. 9 hours

3. If a car travels at 55 miles per hour, how many miles will it travel in five hours?

   A. 60 miles

   B. 250 miles

   C. 275 miles

   D. 300 miles

4. The volume of a cylinder is found by using the formula $V = \pi r^2 h$. How do the volumes of cylinder A and cylinder B compare?

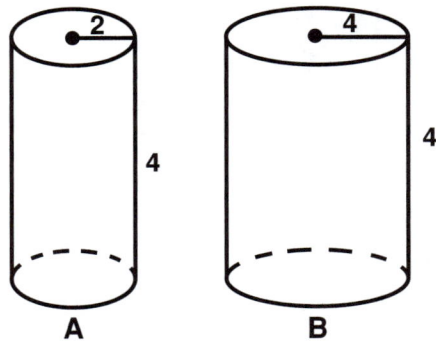

A. The volume of cylinder A is larger.

B. The volume of cylinder B is larger.

C. It is not possible to compare the volumes.

D. The volumes of cylinder A and cylinder B are the same.

5. A good estimate for the weight of a can of soup would be

A. 8 ounces.

B. 8 pounds.

C. 8 gallons.

D. 8 tons.

6. To determine the length of one of your fingers, which is the most appropriate unit of measure?

A. decimeters

B. millimeters

C. centimeters

D. meters

7. What is the approximate area of a circle with a radius of 10 cm?
   (Use $A = \pi r^2$ and $\pi = 3.14$)

   A. 3.14 cm^2

   B. 31.4 cm^2

   C. 314 cm^2

   D. 3140 cm^2

8. Given that water freezes at 0°C and boils at 100°C, what would be the most likely temperature of a person who is not running a fever? (*F* is for Fahrenheit.)

   $$F = \frac{9}{5} \cdot C + 32$$

   A. between 0°C and 10°C

   B. between 10°C and 20°C

   C. between 30°C and 40°C

   D. between 80°C and 90°C

9. Rounding off to the nearest cubic centimeter, estimate the volume of the box.

   A. 17 cm^3

   B. 13 cm^3

   C. 36 cm^3

   D. 127 cm^3

   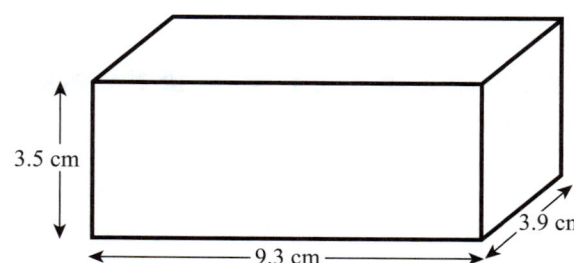

10. Casey lives four miles from George. How many feet is this?

A. 120 feet

B. 4,000 feet

C. 21,120 feet

D. 24,000 feet

Chapter 5 Answers

Let's Review 8: Measurement

1. D
An automobile weighs about two tons.

2. A
A pencil is best measured in inches.

3. C
A typical three-year-old child weighs about 30 pounds.

4. B
There are 16 ounces in a pound. To solve this problem, multiply 16 by 10.

5. B
To solve this problem, multiply 50 by 6. The answer is 300.

6. C
The most likely temperature for the inside of a refrigerator is 45°F, since the temperature is above freezing.

Let's Review 9: Metric Measurement

1. B

Kilometers are best used to measure the distance from one town to another.

2. D

The smallest unit of measurement listed—answer choice D, millimeter—is the best answer.

3. A

The mass of a carton of milk is best measured in grams.

4. D

A refrigerator's mass is best measured in kilograms.

5. C

There are 1,000 grams in a kilogram. Therefore 1,200 grams is a little over 1 kilogram. Answer choice C is correct.

Let's Review 10: Time, Area, Volume

1. D

When you multiply 10 by 4 by 3, you get 120. Metric volume is expressed as cm³ (centimeters cubed or cubic centimeters).

2. D

To solve this problem, you need to find the area of a circle with a radius of 8 feet. If you substitute this into the formula $A = \pi r^2$, the answer is approximately 201 ft².

3. A

When you multiply the length of the square (4 inches) by the height (4 inches), you get 16 in.², the area of the square.

4. B

The time from 9:00 A.M. to 5:00 P.M. is eight hours. When you deduct 45 minutes from 8 hours, you get $7\frac{1}{4}$ or 7.25, the length of the seminar.

Chapter 5 Review

1. C

The box is a cube. Its length, width, and height measure 5 cm. When you multiply 5 × 5 × 5, you get 125.

2. B

The amount of time between 2:30 P.M. and 11:00 P.M. is 8.5 hours. When you subtract 30 minutes from this time, the amount of time Nancy takes for lunch, you get 8.

3. C

To solve this problem, you need to multiply 55, the number of miles a car travels in an hour, by 5, the number of hours the car travels. The answer is 275 miles.

4. B

The volume of cylinder A is about 50; the volume is cylinder B is about 200.

5. A

A can of soup typically weighs about 8 ounces.

6. C

You would use centimeters to measure one of your fingers.

7. C

When you plug the radius of 10 cm into the formula $A = \pi r^2$, you get approximately 314 cm².

8. C

The body temperature of a person who is not running a fever is about 37°C.
$F = \left(\dfrac{9}{5}\right)(37) + 32 = 98.6°$, which is normal body temperature.

9. D

When you multiply the length, width, and height of the box, the answer is about 127 cm³.

10. C

There are 5,280 feet in a mile. When you multiply this by 4, you get 21,120 feet.

Chapter 6
Measurement and Geometry, Part 2

Standards

- Uses proportions involving similar figures and scale drawings.
- Solves problems involving similar figures and scale drawings.
- Applies ratios to similar geometric figures, as in scale drawings, as well as with mixtures and compound applications.
- Identifies and differentiates between similar and congruent figures and identifies those that have been transformed by rotation, reflection, and translation.
- Graphs points in the coordinate plane, identifies the coordinates, and uses the concept of coordinates in problem situations, such as reading maps.

Some questions on the GHSGT will be about similar figures, which are figures with the same proportion. In this chapter you'll learn how to determine a missing side in similar figures.

Other test questions will be about transformations. When you transform a figure, you move it in a certain way. In this chapter you'll learn about the kinds of transformations on the GHSGT.

Last, you'll learn how to give and identify coordinates for points on a coordinate plane.

Congruent Figures

Figures that are **congruent** are exactly the same size and shape. If you place two congruent figures on top of each other, they will fit exactly. The two figures below are congruent.

These triangles are also congruent:

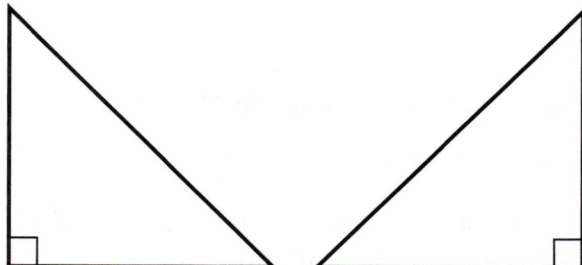

Similar Figures

Figures that are **similar** are not congruent. They have the same shape but not the same size. If figures are similar, their corresponding sides can be written as a proportion because one figure is an enlargement of the other.

These triangles are similar:

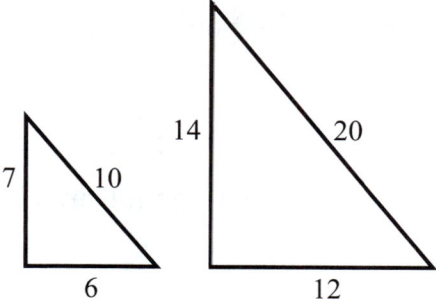

Chapter 6: Measurement and Geometry, Part 2

These rectangles are similar:

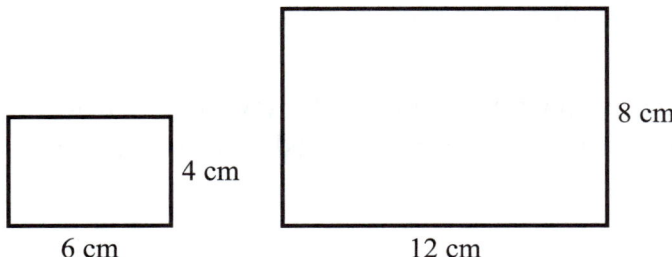

Note that the sides of these rectangles can be written as a proportion that is the same when it is reduced:

$$4 : 6 = 2 : 3$$
$$8 : 12 = 2 : 3$$

Some questions on the GHSGT will ask you to determine the missing side for similar shapes. Read this question:

Find the missing length (*x*) for the pair of similar figures shown.

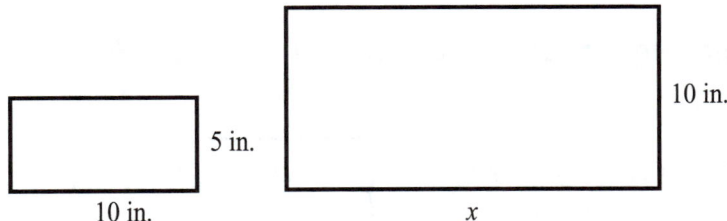

Remember that similar figures have dimensions that can be expressed in the same proportion. The proportion for the first rectangle is shown here:

$$5 : 10 = 1 : 2$$
$$10 : x = 1 : 2$$
$$x = 20 \text{ in.}$$

Let's Review 11: Figure Measurement

Complete each of the following questions. Use the Tip following each question to help you choose the correct answer. When you finish, check your answers with those at the end of Chapter 6.

1. Jamie has a picture that measures 12 inches in width and 24 inches in length. If Jamie enlarges the picture to make a poster 2 feet wide, how long is the poster?

   A. 2 ft

   B. 4 ft

   C. 6 ft

   D. 8 ft

   TIP: The proportion for 12 : 24 is 1 : 2.

2. Find the missing length (*x*) for the pair of similar figures shown.

   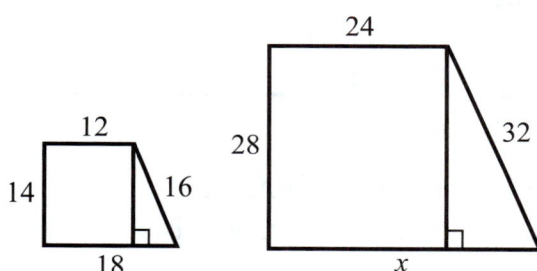

   A. 24

   B. 28

   C. 32

   D. 36

   TIP: The corresponding side on the smaller figure is 18.

3. If the two triangles shown below are similar, find the measure of side x.

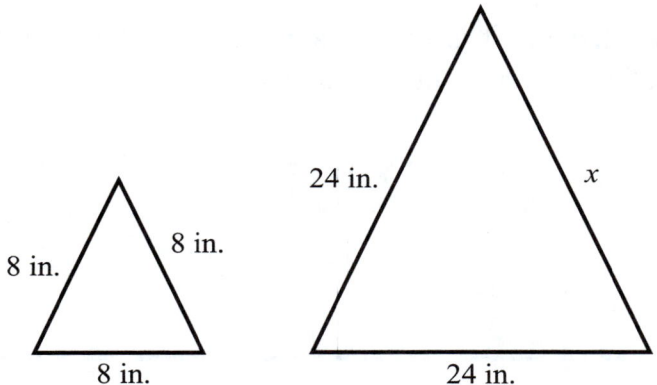

A. 12 in.

B. 24 in.

C. 24 in.

D. 48 in.

TIP: Both triangles are equilateral; this means that all sides are equal.

Transformations

Some test questions will be about transformations, the movement of figures on a coordinate plane. Three types of transformation are tested on the GHSGT: rotation, reflection, and translation.

Rotation

When you rotate a figure, you move it around a fixed point, which is called the center of rotation. A rotation can be large or small. A rotation of 180° is called a half-turn. A rotation of 90° is called a quarter-turn.

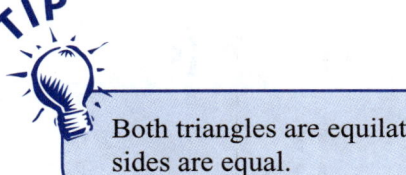

Reflection

When a figure is reflected, it is flipped across a line, which may or may not be visible. A reflection of a figure is a mirror image.

R|Я

Translation

A translation is a slide. A figure that is translated is moved as if you were sliding it in one direction.

R→R

R↘R

Let's Review 12: Figure Transformations

Complete each of the following questions. Use the Tip following each question to help you choose the correct answer. When you finish, check your answers with those at the end of Chapter 6.

1. Study Figures I and II. Which transformation, if any, of Figure I, is shown in Figure II?

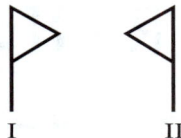

I   II

A. no transformation

B. reflection

C. rotation

D. translation

If you're not sure of the answer, go back and reread the description of each transformation.

2. **Moving a geometric figure around a fixed point is transformation by**

A. inversion.

B. reflection.

C. rotation.

D. translation.

Inversion was not discussed, so you can eliminate answer choice A.

3. **Study Figures I and II. Determine which transformation, if any, of Figure I is shown in Figure II.**

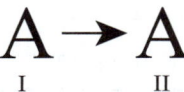

A. rotation

B. reflection

C. translation

D. no transformation

The figure looks as if it has been slid.

The Coordinate Plane

Some test questions will be about coordinate planes. A coordinate plane is a graph with four quadrants, I, II, III, and IV. It has an *x*-axis and a *y*-axis. The *x*-axis is a horizontal line and the *y*-axis is vertical. Look at the coordinate plane shown below. Find the *x*-axis and the *y*-axis and look at the different quadrants.

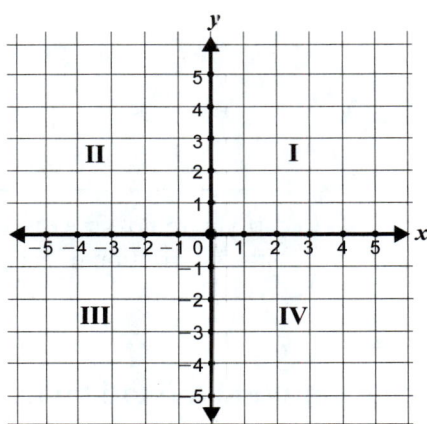

On the GHSGT you might be asked to find the coordinates of a point on the coordinate plane. To find coordinates of a point, move along the x-axis first. If the number of the first coordinate is positive, move to the right. If it's negative, move to the left. Then move along the y-axis. If the number is positive, move up. If it's negative, move down. Look at the coordinate grid shown here. Note that the coordinates of point A are (4, 3).

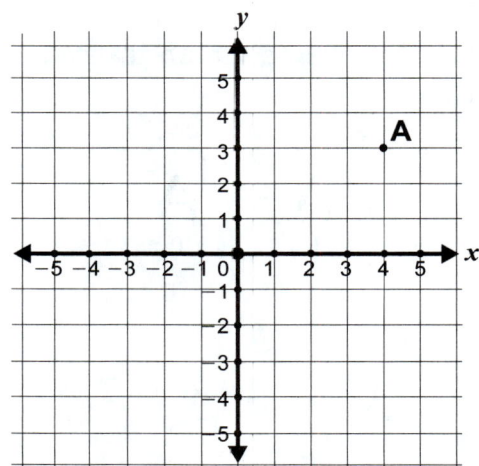

You might also be given coordinates and asked to identify a point. Look at the coordinate plane below. What point is located at (3, –4)?

Chapter 6: Measurement and Geometry, Part 2

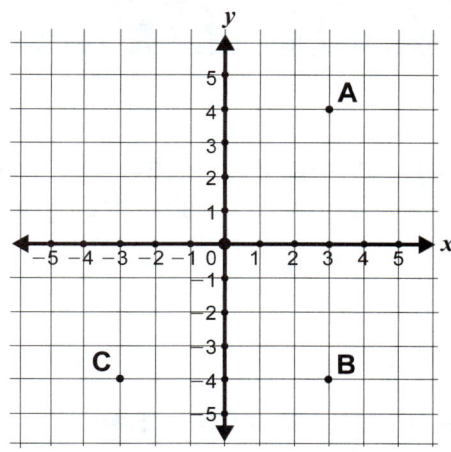

Point B is located at (3, –4).

Other questions on the GHSGT might ask you to interpret a map on a coordinate plane. Look at this map:

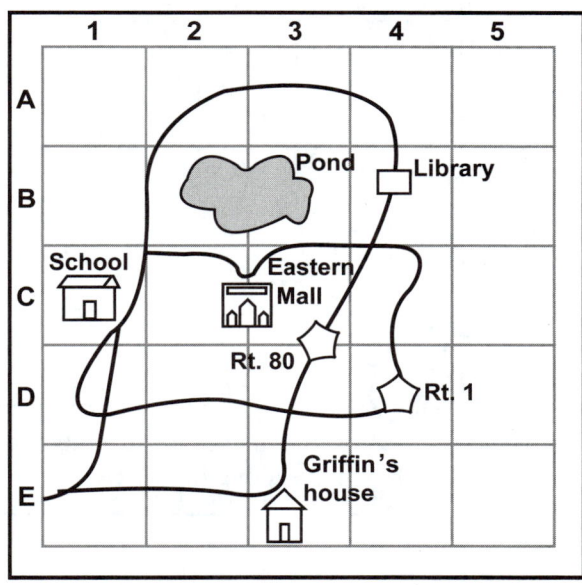

You might be asked to identify a square showing something, such as square 3, E, which shows Griffin's house.

Let's Review 13: Coordinate Plane

Complete each of the following questions. Use the Tip following each question to help you choose the correct answer. When you finish, check your answers with those at the end of Chapter 6.

1. Give the coordinates of point P on the graph.

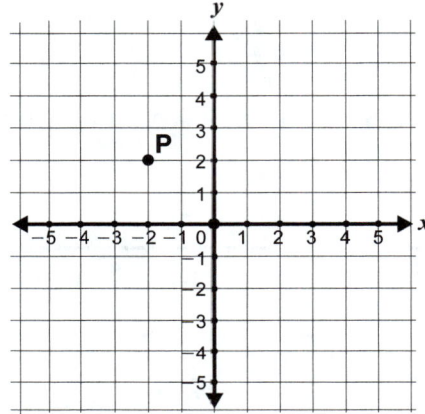

A. (2, –2)

B. (0, 2)

C. (–2, 2)

D. (1, –2)

TIP Remember to move along the *x*-axis first. Then move along the *y*-axis.

Chapter 6: Measurement and Geometry, Part 2

2. Which of the following indicates the square where the airport is located?

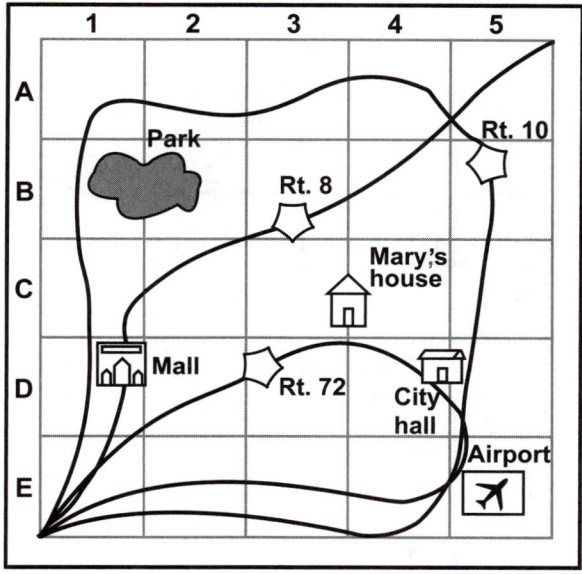

A. 4, D

B. 4, E

C. 5, D

D. 5, E

TIP Find the airport. Then find its coordinates.

3. Which point on the graph has the coordinates (–4, –2)?

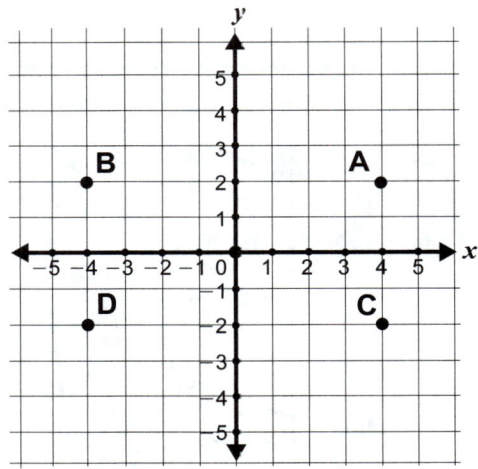

- A. point A
- B. point B
- C. point C
- D. point D

TIP Remember to move along the *x*-axis first.

Chapter 6 Review

Complete each of the following practice problems. Check your answers at the end of this chapter. Be sure to read the answer explanations!

1. Study Figures I and II. Determine which transformation, if any, of Figure I is shown in Figure II.

- A. rotation
- B. reflection
- C. translation
- D. no transformation

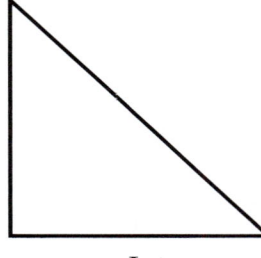
I

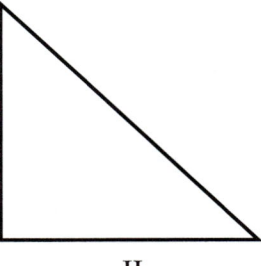
II

2. Enrico has a picture that measures 5 inches in height and 7 inches in length. If Enrico enlarges the picture so that the height is 15 inches, what is its length?

A. 7 in.

B. 10 in.

C. 14 in.

D. 21 in.

3. Study Figures I and II. Determine which transformation, if any, of Figure I is shown in Figure II.

A. rotation

B. reflection

C. translation

D. no transformation

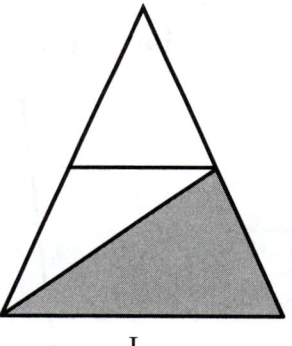
I

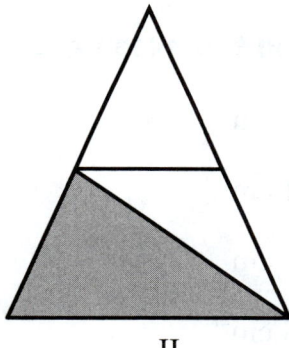
II

4. Give the coordinates of point A on the graph.

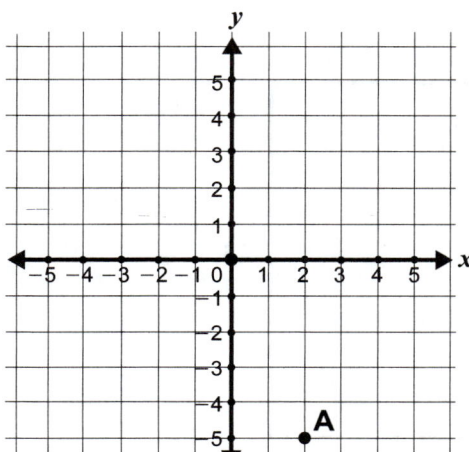

A. (–5, 2)

B. (2, –4)

C. (2, –5)

D. (1, –5)

5. Which point shown on the graph has the coordinates (4, –1)?

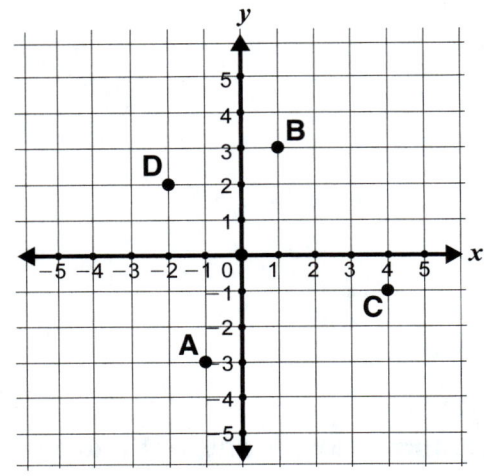

A. point A

B. point B

C. point C

D. point D

6. Find the missing length (x) for the pair of similar figures shown.

A. 16 cm

B. 24 cm

C. 36 cm

D. 48 cm

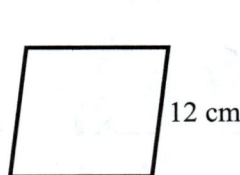
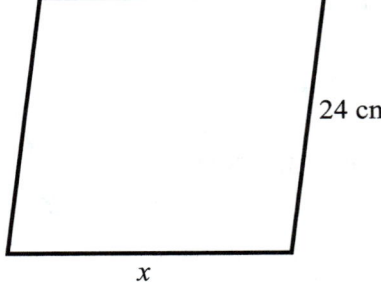

7. Look at the similar figures shown below.

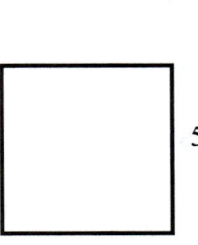
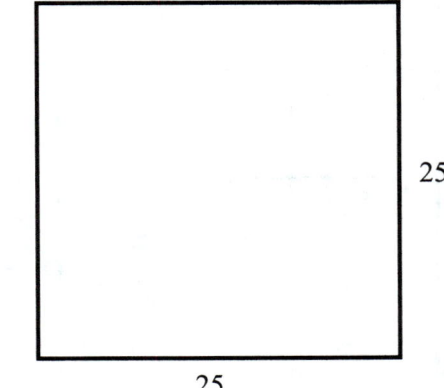

Which ratio gives the correct value for any two sides within each square?

A. 1:1

B. 1:5

C. 5:1

D. 5:20

8. Give the coordinates of point C on the graph.

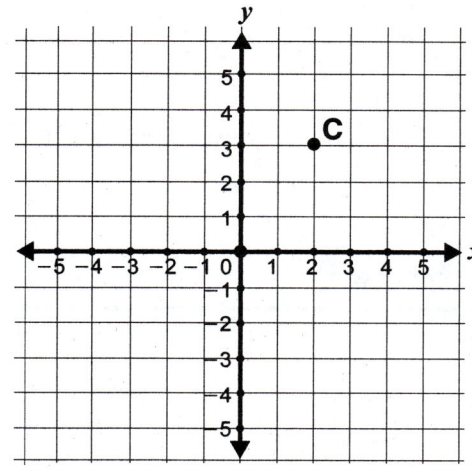

A. (2, 3)

B. (3, 2)

C. (1, 3)

D. (3, 1)

Chapter 6 Answers

Let's Review 11: Figure Measurement

1. B

The ratio for the picture and poster is 1:2. If the poster is 2 feet wide, it is 4 feet long.

2. D

The corresponding side on the smaller figure is 18 and the ratio for the two figures is 1:2, so the correct answer is 36.

3. C

The ratio for these triangles is 1:3 and the corresponding side of the smaller triangle is 8, so the correct answer is 24.

Let's Review 12: Figure Transformations

1. B

The figure shows a reflection or a mirror image.

2. C

The definition in this question is for a rotation.

3. C

The letter is slid from one place to another. This is a translation.

Let's Review 13: Coordinate Plane

1. C

To get to point P, you have to move to -2 on the x-axis and 2 on the y-axis.

2. D
The airport is located in block 5, E.

3. D
If you move to −4 on the x-axis and −2 on the y-axis, you will get to point D.

Chapter 6 Review

1. C
This figure is slid.

2. D
If the height is enlarged from 5 to 15, it is three times its original size. If the length is 7, it will be 21 if it is enlarged.

3. B
The figure is reflected. It is a mirror image.

4. C
The coordinates for point A are (2, −5).

5. C
Point C has the coordinates (4, −1).

6. A
The two figures are similar parallelograms. If the length of the first figure is 8, the length of the second is 16.

7. A

All sides are equal on a square; therefore, the ratio is 1:1.

8. A

The coordinates for point C are (2, 3).

Chapter 7
Measurement and Geometry, Part 3

Standards

- Finds the perimeter and area of plane figures (such as polygons, circles, composite figures) and surface area or volume of simple solids (such as rectangular prisms, pyramids, cylinders, cones, spheres).

- Identifies lines, angles, circles, polygons, cylinders, cones, rectangular solids, and spheres in everyday objects.

- Applies geometric properties—such as the sum of the angles of a polygon property, percent of area of a circle determined by the central angle measure in a pie chart, or parallel sides and angle relations for parallelograms—to practical drawings.

- Draws and measures angles; determines the number of degrees in the interior angles of geometric figures, such as right and straight angles, circles, triangles, and quadrilaterals; and classifies angles (right, acute, obtuse, complementary, supplementary) and triangles (right, acute, obtuse, scalene, isosceles, and equilateral).

- Uses the Pythagorean theorem to solve problems (includes selecting appropriate tools such as the calculator).

In this chapter, you'll expand your knowledge of Measurement and Geometry and learn how to find the perimeter of plane figures. You'll learn about angles and triangles, and you'll learn how to use the Pythagorean theorem to find the missing length of a side of a right triangle.

Perimeter

Perimeter is the distance around a plane figure. A **plane figure** is a flat, closed figure. The following chart lists some common plane figures. Perimeter is commonly measured in inches, feet, centimeters, and meters. The perimeter of a circle is called the **circumference**.

PLANE FIGURES

Figure	Description	Illustration
Rectangle	Has four right angles and four sides; sides across from each other are parallel and equal in length. The sum of the angles in a rectangle is 360°.	
Square	Has four right angles and four sides equal in length. The sum of the angles in a square is 360°.	
Rhombus	Has four equal sides, but no right angles. The sum of the angles in a rhombus is 360°.	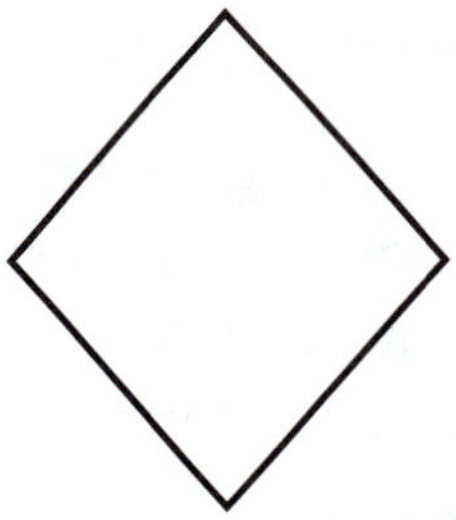
Parallelogram	Has four sides and two pairs of equal, opposite, parallel sides and no right angles. The sum of the angles in a parallelogram is 360°.	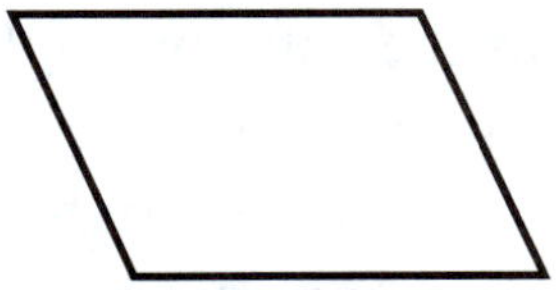
Trapezoid	Has four sides and one pair of parallel sides; may or may not have a right angle. The sum of the angles in a trapezoid is 360°.	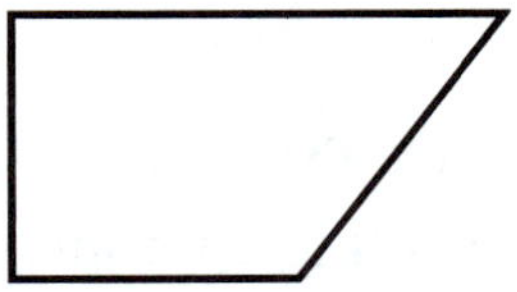
Triangle	Has three sides; length of sides can vary; may or may not have a right angle. The sum of the angles in a triangle is 180°.	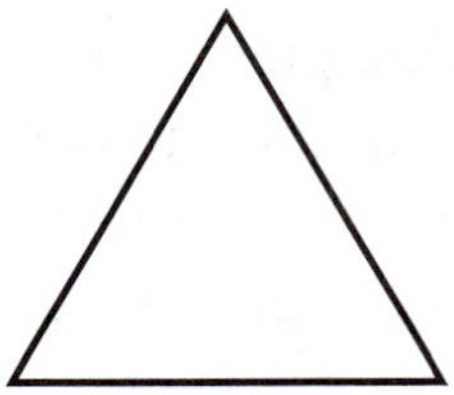

Circle	Has no sides. The sum of the degrees in a circle is 360°.	
Pentagon	Has five sides that may or may not be equal. The sum of the angles in a pentagon is 540°. In a regular pentagon, each angle measures 108°.	
Hexagon	Has six sides that may or may not be equal. The sum of the angles in a hexagon is 720°. Note that a regular hexagon has equal sides. In a regular hexagon, each angle is 120°.	
Octagon	Has eight sides that may or not be equal. The sum of the angles in an octagon is 1,080°. Note that a regular octagon has equal sides. Each angle in a regular octagon measures 135°.	

Find the perimeter of the hexagon shown:

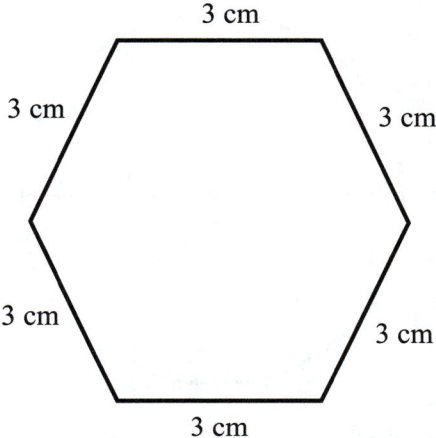

To find the perimeter, add all the sides, $3 + 3 + 3 + 3 + 3 + 3 = 18$ cm, or $6 \times 3 = 18$ cm.

On the GHSGT, you might be asked a question like this about perimeter:

A regular hexagon has a perimeter of 36 inches. What is the length of each of its sides?

To answer this question, you need to know that a hexagon has six sides. If the hexagon is regular, the sides are equal in the length. If the perimeter is 36, you can divide it by 6 (the number of sides) to find the length of the sides. The sides are 6 inches long.

You just learned that the perimeter of a circle is called the **circumference**. Use this formula to find a circle's circumference: $C = \pi \times diameter$. Look at this circle:

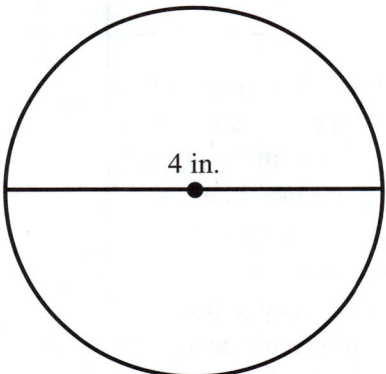

To find the circumference, multiply π (3.14) by the diameter, 4. Then round to the nearest whole number. The circumference is approximately 13 inches.

Sometimes only the radius of a circle is given. When this happens, you have to double the radius, since the diameter is twice the length of the radius. Look at this circle:

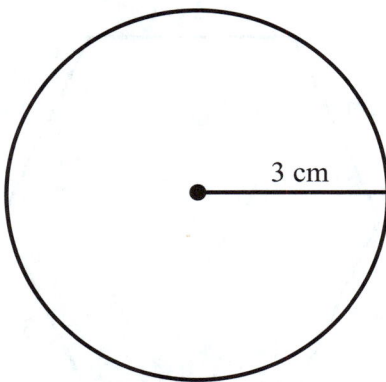

To find the circumference of this circle, first double the radius: $3 \times 2 = 6$. Then plug the numbers into the formula $C = \pi \times diameter$.

$$C \approx (3.14) \times 6$$
$$C \approx 19 \, cm$$

Chapter 7: Measurement and Geometry, Part 3

Let's Review 14: Perimeter

Complete each of the following questions. Use the Tip following each question to help you choose the correct answer. When you finish, check your answers with those at the end of Chapter 7.

1. The perimeters of the two triangles are equal. What is the value of x?

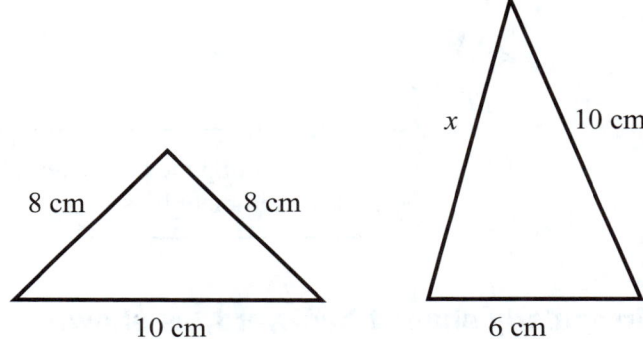

A. 4 cm

B. 6 cm

C. 8 cm

D. 10 cm

TIP: To solve this problem, add the sides of the first triangle, and then subtract the total of the two sides on the second triangle from the perimeter of the first.

2. A hexagon has a perimeter of 25. Five of its sides are 3, 3, 4, 6, and 6. What is the length of the remaining side?

A. 2

B. 3

C. 4

D. 5

TIP: Add the five sides. Then subtract this value from 25.

3. The regular hexagon shown here has the same perimeter as a square with a side of 18 inches. How long is each side of the hexagon?

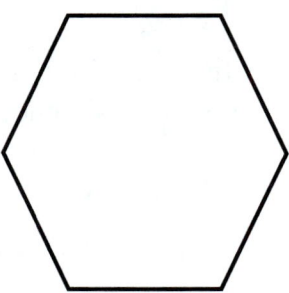

A. 10 inches

B. 12 inches

C. 24 inches

D. 72 inches

Begin by finding the perimeter of the square. Then divide this number by six.

4. Jeffrey has an irregularly shaped backyard, as shown.

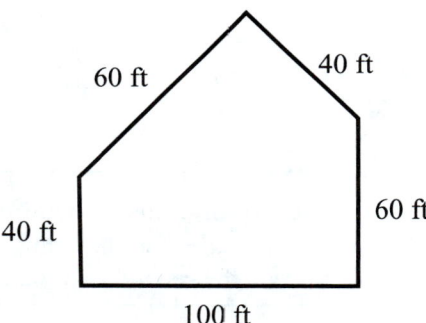

What is the perimeter of his backyard?

A. 170 feet

B. 180 feet

C. 200 feet

D. 300 feet

Add together the sides of Jeffrey's yard to solve this problem.

Lines

A **line segment** is part of a line. It has two **endpoints**, one at each end, to show that that it stops and doesn't keep on going.

A **ray** is also part of a line, but unlike a line segment, it keeps on going in one direction. A ray has only one endpoint.

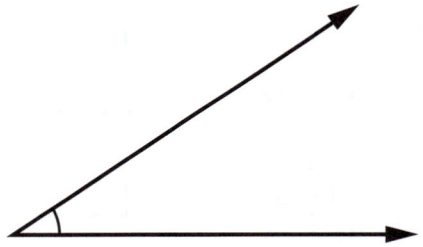

Two rays join together to form an **angle**. The place where they join is called the **vertex**.

A **line** has an arrow on both ends to show that it keeps going.

Lines that do not intersect are called **parallel lines**. Strings on a guitar are parallel, like the lines shown here:

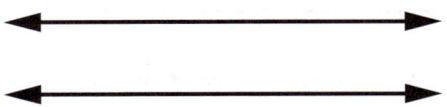

Lines that intersect to form right angles are called **perpendicular lines**. The place where the lines intersect is called the **point of intersection**.

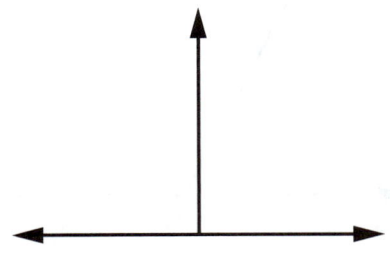

Angles

Angles can be classified by their degree. The following chart shows some angles that you need to know for the GHSGT.

ANGLES

Angle	Description	Illustration
Acute angle	Less than 90 degrees	
Right angle	Exactly 90 degrees	
Obtuse angle	Greater than 90 degrees and less than 180 degrees	
Straight angle	Exactly 180 degrees	
Reflex angle	Greater than 180 degrees and less than 360 degrees	

Angle Relationships

The sign ∠ stands for the word angle. **Adjacent angles** are angles that share a side. In the following illustration, ∠ADC and ∠CDB are adjacent angles.

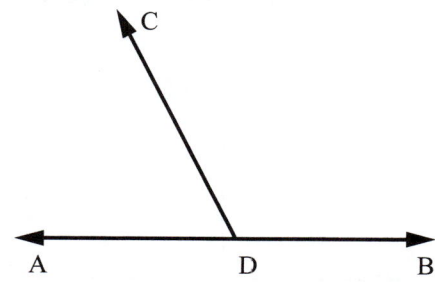

Two angles that add up to 90° are called **complementary angles**. Two angles that add up to 180° are called **supplementary angles**. The angles shown above are supplementary. ∠ADC measures 60° and ∠CDB measures 120°.

Vertical angles, angles across from each other, are always equal. In the illustration below, angles A and C are equal and angles B and D are equal.

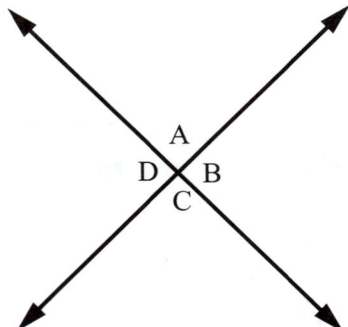

Sometimes parallel lines are intersected by a line. This intersecting line is called a **transversal**, and it creates eight angles, four of which are acute and four of which are obtuse. Look at these parallel lines intersected by a transversal:

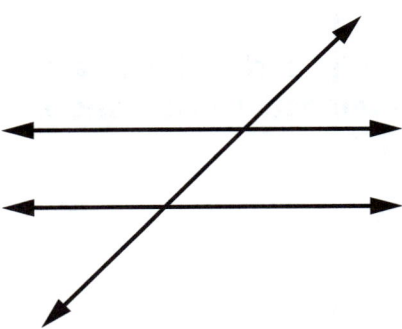

When parallel lines are intersected by a transversal, the four acute angles are always equal and the four obtuse angles are always equal. Each acute and obtuse angle form supplementary angles, whose sum is 180°.

Let's Review 15: Lines and Angles

Complete each of the following questions. Use the Tip following each question to help you choose the correct answer. When you finish, check your answers with those at the end of Chapter 7.

1. Sarah's yard and her neighbor's yard form complementary angles. What is the sum of the degree of measures of the two yards?

   A. 80°

   B. 90°

   C. 120°

   D. 180°

   If you're not sure the total measure of complementary angles, reread the information in the previous section of this chapter.

2. Two streets in Josh's neighborhood run next to each other in the same direction but do not intersect. These streets are an example of what kind of lines?

   A. perpendicular

   B. adjacent

   C. supplementary

   D. parallel

   Try to remember the name for two horizontal lines that do not intersect.

Triangles

A **triangle** is a plane figure with three sides. Each of the three points on a triangle is called a vertex. You read earlier in this chapter that the sum of the angles in a triangle is 180°. The chart below describes types of triangles.

TRIANGLES

Triangle	Description	Illustration
Equilateral triangle	Has three equal sides; each angle is 60°.	
Isosceles triangle	Has at least two equal sides and two equal angles. For example, an isosceles triangle might have angles measuring 80°−50°−50°.	80°, 50°, 50°
Scalene triangle	Has no equal sides and no equal angles.	
Right triangle	Has one right angle. The side opposite the right angle is called the **hypotenuse**. The other two sides are called **legs**.	

Pythagorean Theorem

The Pythagorean theorem is a formula used to find the length of one side of a right triangle when you know the length of the other two. The formula is $a^2 + b^2 = c^2$.

Look at this triangle:

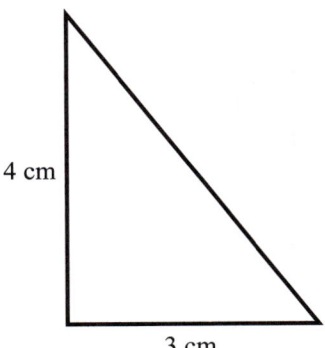

4 cm

3 cm

To find the length of the hypotenuse in this triangle, substitute 3 cm and 4 cm into the formula $a^2 + b^2 = c^2$:

$$3^2 + 4^2 = c^2$$
$$9 + 16 = 25$$

Then find the square root of 25: 5.

The length of the hypotenuse is 5 cm.

If you are given the length of side C, the hypotenuse, but are missing the length of either leg A or leg B, you can still use the Pythagorean theorem to find the missing side. Look at this triangle:

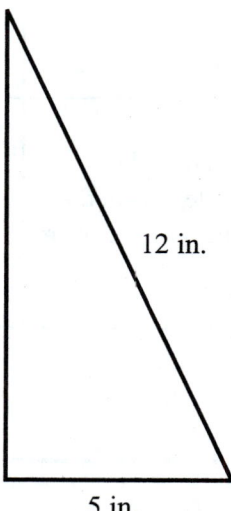

$$c^2 - a^2 = b^2$$
$$12^2 - 5^2 = b^2$$
$$b^2 = 144 - 25$$
$$b^2 = 119$$

$\sqrt{119}$ = approximately 11 in.

Chapter 7: Measurement and Geometry, Part 3   137

You can find the square root easily by using your calculator. Once you find the length of the missing side by using the Pythagorean theorem, simply enter the square root sign.

Circles

Some questions on the GHSGT will ask you about circles. Angles can be formed inside a circle when two radii meet and share a vertex. When this happens, the angles are called **central angles**. Look at the circle shown below. It has two central angles.

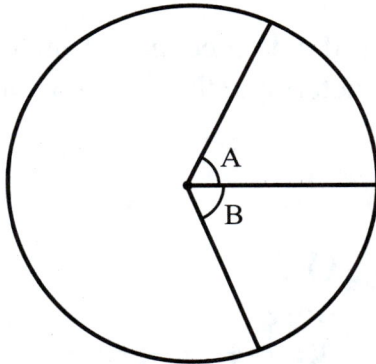

To determine the length of the number of central angles that will fit into a circle, remember that a circle is 360°, and that half of a circle is 180°. For example, in the circle below, you can figure out the measure of the missing measurement by subtracting the sum of the measures of the other angles from 360°.

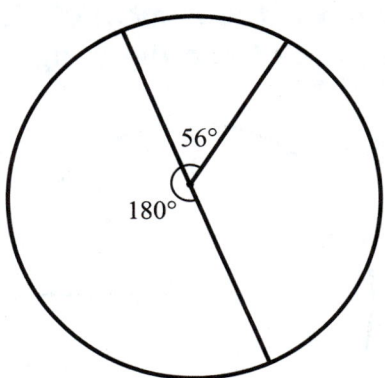

Let's Review 16: Triangles and Circles

Complete each of the following questions. Use the Tip following each question to help you choose the correct answer. When you finish, check your answers with those at the end of Chapter 7.

1.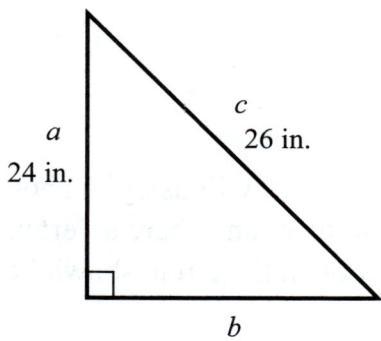

    In this drawing, the length of side *a* equals 24 inches. The length of side *c* is 26 inches. Which formula would determine the length of side *b*?

    A. $a^2 + c^2 = b^2$

    B. $a^2 + b^2 = c^2$

    C. $a^2 - b^2 = c^2$

    D. $c^2 - a^2 = b^2$

    TIP
    Remember that you need to put side B on one side of the equation.

2. **Beth wants to make a design with a circle divided into pie-shaped pieces of equal size. What is the number of pieces Beth can have if she wants the central angles to be right angles?**

    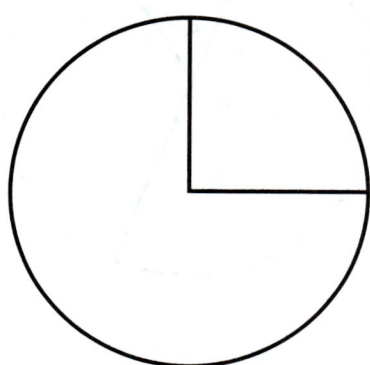

A. 2

B. 3

C. 4

D. 5

TIP: The pie-shaped piece is one-quarter of the size of the circle.

Chapter 7 Review

Complete each of the following practice problems. Check your answers at the end of this chapter. Be sure to read the answer explanations!

1. **The following hexagon is regular. What is the measure of each of its angles?**

   A. 90°

   B. 108°

   C. 120°

   D. 180°

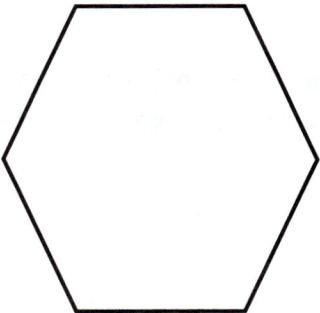

2. **An irregular octagon has a perimeter of 64. Seven of its sides measure 4, 4, 9, 8, 8, 5, and 10. What is the length of the remaining side?**

   A. 10

   B. 12

   C. 14

   D. 16

3. **What is the measure of ∠DHE?**

   A. 60°

   B. 90°

   C. 180°

   D. 360°

   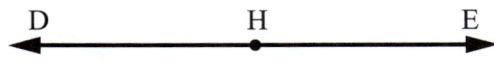

4. **Elm Street and Maple Street are parallel to each other. Walnut Street crosses Elm Street and Maple Street. What is the measure of angle A?**

   A. 20°

   B. 25°

   C. 75°

   D. 80°

   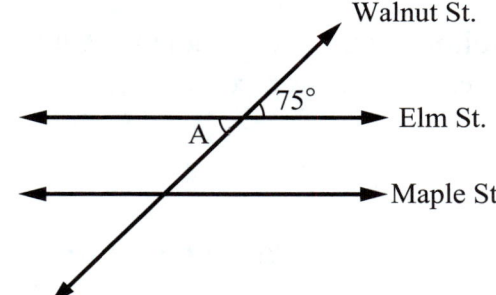

5. **Two streets in Terry's neighborhood intersect and form four right angles. These streets are an example of what kind of lines?**

   A. perpendicular

   B. adjacent

   C. supplementary

   D. parallel

6. **Look at the following right triangle. What is the approximate length of side c?**

   A. 4 cm

   B. 6 cm

   C. 8 cm

   D. 16 cm

   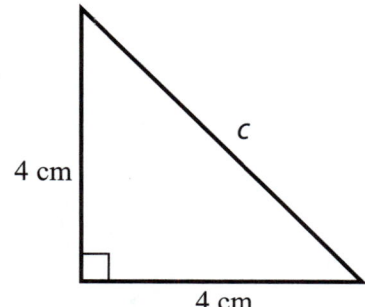

Chapter 7 Answers

Let's Review 14: Perimeter

1. D

The perimeter of the first triangle is 26. To get x, subtract the sum of the two sides of the second triangle from this number.

2. B

Add the five sides of the hexagon, and then subtract this number from the perimeter.

3. B

The perimeter of the square is $18 + 18 + 18 + 18$. Then divide this number by six to get the perimeter of the hexagon.

4. D

Add up all of the sides. The perimeter of his backyard is 300 feet.

Let's Review 15: Triangles and Circles

1. B

The sum of complementary angles is 90 degrees.

2. D

Lines that do not intersect are parallel.

Let's Review 16: Triangles and Circles

1. D

Since you know the measure of sides a and c, you would subtract a^2 from c^2, then take the square root.

2. C

Beth's design will be $\frac{1}{4}$ of the circle.

Chapter 7 Review

1. C

A regular hexagon has angles that measure 120 degrees.

2. D

To find this answer, add up all of the sides of the octagon and subtract from the perimeter.

3. C

The measure of a straight angle is 180 degrees.

4. C

Angle A is congruent to the angle labeled 75 degrees.

5. A

Two lines that intersect to form four right angles are perpendicular.

6. B

If you use the Pythagorean theorem, you'll see that the length of side c is approximately 6 cm.

Chapter 8
Algebra, Part 1

Standards

- Simplifies expressions with and without grouping symbols.
- Evaluates simple algebraic expressions.
- Solves simple equations, including addition, subtraction, multiplication, division, proportions, and two-step equations.
- Translates words into simple algebraic expressions and equations.
- Substitutes known values in formulas and solves problems with formulas.

For some questions on the GHSGT, you will have to choose the correct way to simplify an algebraic expression. For others, you will have to choose the correctly worked solution to an algebraic expression. You'll learn about algebraic expressions in this chapter.

Other test questions will ask you about equations, in which one value is equal to another. You will learn how to write and solve basic equations in this chapter.

Simplifying Expressions

In algebra, letters often stand for numbers that need to be determined. These letters are called **variables**. Keep in mind that any variable, such as x or y, has a 1 before it, even though the 1 is not written. In other words, $x = 1x$ and $y = 1y$.

Any number in front of a variable means that the variable will be multiplied by that number. For example, $2x$ means that x will be multiplied by 2.

Parentheses often are used in algebraic expressions. According to the **distributive property**, multiplication is performed, or distributed, before addition and subtraction. Look at this expression:

$$2(x + y) = 2x + 2y$$

On the GHSGT, you will be asked to simplify simple expressions, such as the one shown here:

$$4(3x) + x$$

Begin simplifying this expression by multiplying:

$$12x + x$$

Now simplify it even further by adding. Remember that x is the same as $1x$.

$$13x$$

Other test questions will ask you to choose the correct expression based on a given situation. Read this question:

Mabel bought 12 pencils for 20 cents each, 6 pens for 50 cents each, 2 erasers for 50 cents each, and 5 sheets of construction paper for 20 cents each. Which expression would enable Mabel to find out how much money she spent?

To write an expression for this question, you need to add items with the same cost. For example, pencils and construction paper cost 20 cents each. So, you can add the 12 pencils and 5 sheets of construction paper and multiply them by 20, the cost.

$$17(20)$$

You can also add the pens and erasers, since they both cost 50 cents each.

$$8(50)$$

Next you need to write an expression in which the 20-cent items and the 50-cent items are added:

$$17(20) + 8(50)$$

Although you could use a calculator to quickly evaluate or solve this expression, for this type of question on the GHSGT, you would simply choose the correct expression.

Let's Review 17: Simplifying Expressions

Complete each of the following questions. Use the Tip following each question to help you choose the correct answer. When you finish, check your answers with those at the end of Chapter 8.

1. **A sporting goods store had 52 sweatshirts at the beginning of a sale. If y represents the number of sweatshirts sold during the sale, which expression shows the number of sweatshirts remaining?**

    A. $y - 52$

    B. $52(y)$

    C. $52 - y$

    D. $y + 52$

    TIP Remember that the store had 52 sweatshirts before the sale, so the amount sold would have to be subtracted from this amount.

2. **Simplify the following expression, if possible. ($x \neq 0$)**

    $$\frac{2x}{x^2}$$

    A. x

    B. $2x$

    C. $\frac{2}{x}$

    D. 2

    TIP You can cancel one x in both the numerator and denominator of this fraction.

3. **Which statement is represented by the algebraic expression 5 + x?**

   A. Melanie has lived in her house for five years. How long will she have lived there in x years?

   B. Ashton has lived in his house for x years. How many years did he live there five years ago?

   C. Peter has lived in his house x more years than Chris. How long has Chris lived in his house?

   D. Adrienne has lived in her house for five years. How many years did she live there x years ago?

> **TIP**
> Remember that 5 is known and x is the variable, which means it is the unknown quantity.

Evaluating Expressions

To evaluate expressions on the GHSGT, you may have to substitute a value for a variable and then find the value of the expression. For example, look at this expression:

$$2x + 3$$

If a test question tells you to substitute 4 for x, you would evaluate the expression this way:

$$2(4) + 3$$
$$8 + 3$$
$$11$$

Other test questions will ask you to choose a situation that represents an algebraic expression. For example, read this question:

Abraham uses the expression $7x + 10.5y$ to determine the amount he earns at a pay rate of seven dollars an hour plus time and a half for overtime. One week he worked 40 hours, plus two hours of overtime. Which expression best represents his total pay for the week?

The information in this problem gives you the values to substitute for x and y. Substitute 40 for x, the number of regular hours Abraham worked, and substitute 2 for y,

Chapter 8: Algebra, Part 1

the number of overtime hours Abraham worked. This expression would help Abraham determine how much money he would earn in a week.

$$7(40) + 10.5(2)$$

Let's Review 18: Evaluating Expressions

Complete each of the following questions. Use the Tip following each question to help you choose the correct answer. When you finish, check your answers with those at the end of Chapter 8.

1. The number of plain white straws Cara has is shown by the expression $3x + 4$, with x representing her striped straws. If Cara has 10 striped straws, how many plain white straws does she have?

   A. 16
   B. 30
   C. 34
   D. 70

   TIP Evaluate the expression like this: $3(10) + 4$.

2. Find the numerical value of $3x^2 + x$, when $x = 8$.

   A. 34
   B. 192
   C. 200
   D. 292

   TIP Evaluate this expression like this: $3(8^2) + 8$.

3. Morgan's age is shown by the expression $a + 3$, where a represents Andrea's age. If Andrea is 9, how old is Morgan?

A. 6

B. 9

C. 12

D. 15

TIP: Add 9 and 3.

Solving Equations

An algebraic **equation** is a statement that says two values are equal. You can spot an equation easily because it has an equal sign, which separates the two sides of the equation. To keep the sides of an equation equal, you must perform the same operation to both sides. For example, look at the equation shown here:

$$t + 45 = 100$$

What do you think t is? To solve this equation, you need to subtract 45 from 100, like this:

$$t + 45 - 45 = 100 - 45$$

Notice that -45 has been added to both sides, but it cancels out 45 on the left side of the equal sign. Now solve the equation.

$$t = 55$$

Let's try one more.

What is the value of x, if $3x + 2 = 20$?

To solve this equation, subtract 2 from each side.

$$3x - 2 + 2 = 20 - 2$$
$$3x = 18$$
$$x = \frac{18}{3}$$
$$x = 6$$

Chapter 8: Algebra, Part 1

Translating Words in Expressions and Equations

Some questions on the GHSGT will give you a problem expressed in words, and you'll be asked to translate this problem into either an expression or equation. For example, you might be given a phrase such as "six reduced by the product of a number and two." You can tell from the wording of the phrase that this is an expression and not an equation. And you can tell by the keywords "reduced" and "product" that the expression will contain a minus sign and a multiplication sign. The words "of a number" refer to the variable. Look at the phrase and the expression that follows:

six reduced by the product of a number and two

$$6 - 2x$$

Solving Problems with Formulas

In Chapter 5, Measurement and Geometry, Part 1, you learned how to solve problems by using a formula. You used formulas to find the area and volume of geometric figures. You may be asked to use other formulas on the GHSGT, and some of these formulas will contain variables and involve an equation. Read this problem.

Sydney drives from his hometown to the beach in eight hours. If he knows his average speed is 50 mph, what is the distance from his hometown to the beach?

(Use the formula $d = rt$)

The variables in this formula stand for distance (d), rate (r), and time (t).

Sydney's average speed, 50 mph, is the rate and 8 hours is the time.

$$d = 50 \times 8$$
$$d = 400$$

Let's Review 19: Solving Equations

Complete each of the following questions. Use the Tip following each question to help you choose the correct answer. When you finish, check your answers with those at the end of Chapter 8.

1. Which operation would be used to solve the equation $3x = 24$?

   A. addition
   B. division
   C. multiplication
   D. subtraction

   TIP When you put x on one side of the equation, what do you do with the 3, multiply or divide?

2. Which of the following algebraic expressions corresponds to "b increased by 3 is equal to $\frac{1}{4}$ of c"?

   A. $b + 3 = \frac{c}{4}$
   B. $3b = \frac{c}{4}$
   C. $b + 3 = 4c$
   D. $3b = \frac{c}{4}$

   TIP You would write the fraction $\frac{1}{4}$ of c as $\frac{c}{4}$.

3. Solve for y, if $\frac{3}{y} = \frac{1}{2}$.

   A. 2
   B. 3
   C. 4
   D. 6

   TIP What fraction with three as a denominator can be reduced to $\frac{1}{2}$?

4. Which of the following algebraic expressions corresponds to "three added to the product of a number and two"?

A. $3 + 2 + x$

B. $3x + 2$

C. $3 + 2x$

D. $x + 6$

TIP: Remember that "of a number" refers to a variable.

5. Determine the approximate volume of a cylinder with a radius of 4 cm and a height of 10 cm. Use 3.14 for π.

$V = \pi r^2 h$

A. 50 cm³

B. 160 cm³

C. 220 cm³

D. 500 cm³

TIP: When you multiply 3.14 by 16, round to the nearest whole number.

Chapter 8 Review

Complete each of the following practice problems. Check your answers at the end of this chapter. Be sure to read the answer explanations!

1. Which of the following algebraic expressions corresponds to "*y* increased by 3 is equal to one-fifth of *z*"?

   A. $3y = \dfrac{z}{5}$

   B. $y + 3 = 5z$

   C. $\dfrac{3}{y} = \dfrac{5}{z}$

   D. $y + 3 = \dfrac{z}{5}$

2. What is the value of *x*, if $3x + 3 = 12$?

   A. 2

   B. 3

   C. 4

   D. 9

3. Cheryl's math homework required her to simplify, if possible, algebraic expressions. She simplified an expression this way, knowing that the teacher said *y* does not equal 0 in this problem:

$$45x + 2 + \dfrac{3y}{3y} = 45x + 2$$

Which statement explains what Cheryl did?

A. She simplified the expression correctly.

B. She canceled correctly, but the answer is incorrect.

C. She should have canceled only the 3's; the answer is incorrect.

D. She did not add the terms in the numerator; the answer is incorrect.

4. **Simplify, if possible:**

   $2 + 2n(3n)$

A. $2 + 5n$

B. $4n(3n)$

C. $2 + 6n^2$

D. $12n$

5. **Find the numerical value of $9x + x$ when $x = 6$.**

A. 9

B. 54

C. 60

D. 64

6. Alicia's backyard is a square, 100 feet on each side. To calculate the amount of topsoil needed to cover her yard, she had to find the area of the yard. She used the formula $A = 100^2$, where 100 represents the length of one side. How many square feet of topsoil does she need to cover her backyard?

   A. 100 ft^2

   B. $1{,}000 \text{ ft}^2$

   C. $10{,}000 \text{ ft}^2$

   D. $100{,}000 \text{ ft}^2$

7. Which operation would be used to solve the equation $x + 5 = 9$?

   A. addition

   B. subtraction

   C. multiplication

   D. division

8. The number of cats at an animal shelter is shown by the expression $2y - 5$, with y representing the number of dogs. If the shelter has 125 dogs, how many cats does it have?

   A. 130

   B. 245

   C. 250

   D. 255

Chapter 8 Answers

Let's Review 17: Simplifying Expressions

1. C

If the store had 52 sweatshirts before the sale and y represents the unknown number of sweatshirts sold during the sale, an expression that could be used to find the number of sweatshirts remaining is $52 - y$.

2. C

The x in the numerator and one x in the denominator cancel out; the remaining fraction is $\frac{2}{x}$.

3. A

To solve this problem, you have to choose the situation represented by the expression $5 + x$. If Melanie has lived in her house for five years and wants to know how long she will have lived there in x years, the expression is $5 + x$.

Let's Review 18: Evaluating Expressions

1. C

To simplify the expression $3x + 4$, put 10 in place of x: $3(10) + 4 = 34$. Cara has 34 plain white straws.

2. C

To evaluate this expression, substitute 8 for x: $3(8)^2 + 3 = 3(64) + 8 = 200$.

3. C

If Andrea is 9 and Morgan is Andrea's age plus 3, Morgan is 12.

Let's Review 19: Solving Equations

1. B
You would solve the equation $3x = 24$ like this: $x = \frac{24}{3}$. The operation you would use is division.

2. A
The phrase "b increased by 3 is equal to $\frac{1}{4}$ of c" is the same as $b + 3 = \frac{c}{4}$.

3. D
The fraction $\frac{3}{6}$ can be reduced to $\frac{1}{2}$.

4. C
The phrase "three added to the product of a number and two" is the same as $3 + 2x$.

5. D
To solve this problem, plug the values into the formula like this: $V = (3.14)4^2 \times 10$

Chapter 8 Review

1. D
The phrase "y increased by 3 is equal to one-fifth of z" is the same as $y + 3 = \frac{z}{5}$.

2. B
If you substitute 3 into the equation, you can find the value of x: $3(3) + 3 = 12$.

3. B
Cheryl canceled correctly, but she should have solved the problem like this:

$$45x + 2 + \frac{3y}{3y} = 45x + 2 + 1$$
$$= 45x + 3$$

4. C

You can simplify $2 + 2n(3n)$ as $2 + 6n^2$.

5. C

If you substitute 6 for x in $9x + x$ and solve, $9(6) + 6 = 54 + 6 = 60$.

6. C

The number 100 squared is 10,000.

7. B

You would use subtraction to solve the problem like this: $x = 9 - 5$.

8. B

If y represents the number of dogs and there are 125 dogs, substitute 125 for y into this equation like this: $2(125) - 5$.

Chapter 9
Algebra, Part 2

Standards

- Identifies and applies mathematics to practical problems requiring direct and inverse proportions.

- Identifies ratio and proportion as they appear in applied situations and solves proportions for missing numbers in applied problems.

- Solves linear inequalities in one variable and graphs the solution set on the number line.

- Graphs a linear equation in two variables.

- Finds the slope and intercepts of a graphed line.

- Solves problems that involve systems of two linear equations in two variables.

Some questions on the GHSGT will involve proportions and inequalities. You learned a little bit about ratios, which are part of proportions, in Chapter 6. You'll expand your knowledge of ratios and proportions in this chapter so you can answer more difficult test questions.

An **inequality** shows two values that may or may not be equal. You'll learn about inequalities in this chapter.

Last, imagine a very steep mountain. This mountain would have a steep slope. Lines on a graph also have a slope, and you'll learn how to find the slope of a line in this chapter.

Proportions

A **proportion** is a statement that says two ratios are equal. On the GHSGT, ratios within proportions are often fractions, and either the numerator or the denominator of one fraction is a variable. Look at this proportion:

$$\frac{x}{6} = \frac{6}{12}$$

To solve for x in this proportion, cross multiply:

$$12x = 36$$

Then put the variable on one side:

$$x = \frac{36}{12}$$
$$x = 3$$

Some problems on the GHSGT will ask you to set up a proportion. Like many questions on this test, some questions about proportions will involve real-world situations. For example, read this question:

Megan can read 50 pages of a book in 70 minutes. How many pages can Megan read in 120 minutes? Round to the nearest whole number.

You need to set up a proportion to solve this problem:

$$\frac{50}{70} = \frac{x}{120}$$

Then cross multiply:

$$70x = 6{,}000$$

Then put only x on one side of the equation:

$$x \approx 86$$

Megan can read about 86 pages in 120 minutes.

Chapter 9: Algebra, Part 2

For other test questions, you will have to choose a fraction that expresses the correct ratio. For example, read this question:

Suzette takes piano lessons once a week. During the last two months (eight weeks) she attended six lessons but missed two. What is the ratio of the number of piano lessons she attended to the total number of lessons she was supposed to attend?

To determine this ratio, use the total number of lessons (8) as the denominator and the number of lessons Suzette attended (6) as the numerator. The ratio is $\frac{6}{8}$ or $\frac{3}{4}$.

Let's Review 20: Proportions

Complete each of the following questions. Use the Tip following each question to help you choose the correct answer. When you finish, check your answers with those at the end of Chapter 9.

1. **Four oranges sell for $0.80. How much would it cost to purchase eight oranges?**

    A. $1.00

    B. $1.20

    C. $1.40

    D. $1.60

    TIP: Set up the proportion like this: $\frac{4}{.80} = \frac{8}{x}$.

2. A gym teacher takes inventory and discovers that the school has three soccer balls for every seven basketballs. Which proportion should the gym teacher use to determine the number of soccer balls (S) the school will have if it has 21 basketballs?

A. $\dfrac{21}{G} = \dfrac{3}{7}$

B. $\dfrac{G}{7} = \dfrac{3}{21}$

C. $\dfrac{3}{7} = \dfrac{S}{21}$

D. $\dfrac{3}{7} = \dfrac{21}{S}$

TIP One of the fractions should be $\dfrac{3}{7}$, the number of soccer balls compared to the number of basketballs.

3. In a supply closet, there are blue tablets (*b*) and green tablets (*g*). Which fraction expresses the ratio of blue tablets to the total number of tablets?

A. $\dfrac{1}{b}$

B. $\dfrac{b}{(b + g)}$

C. $\dfrac{1}{(b + g)}$

D. $\dfrac{b}{g}$

TIP Be careful here. The denominator should be to *total* number of tablets.

Linear Inequalities

An **inequality** links two expressions that may or may not be equal. A **linear inequality** is an inequality that, when graphed, forms a straight line. You may be asked questions about both inequalities and linear inequalities. Questions about linear inequalities usually are graphed on a number line on the GHSGT.

Inequalities use these signs:

> greater than

< less than

≥ greater than or equal to

≤ less than or equal to

≠ not equal to

Questions about inequalities use words like these: *greater than, less than, between, at least,* and *at most*.

Solving an inequality is very similar to solving an equation. Look at this inequality:

$$3y > 21$$

Put *y* on a side by itself:

$$y > \frac{21}{3}$$
$$y > 7$$

y can be any number that is greater than 7.

Now read this statement and look at the graph describing it.

If it is 0°C or colder, school will be delayed in the morning.

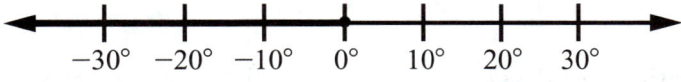

Note that the zero and numbers lower than zero are highlighted on the number line.

Linear Equations

A **linear equation** is an equation that forms a straight line when graphed. When you graph an equation, you substitute values for variables, which are unknown quantities. On the GHSGT, linear equations usually have more than one variable. Look at the next equation:

$$x + y = 4$$

If you listed the numbers that could be substituted for the variables x and y, your list might look like this:

x	y
0	4
1	3
2	2
3	1
4	0

If you graphed these numbers and drew a line, it would look like this:

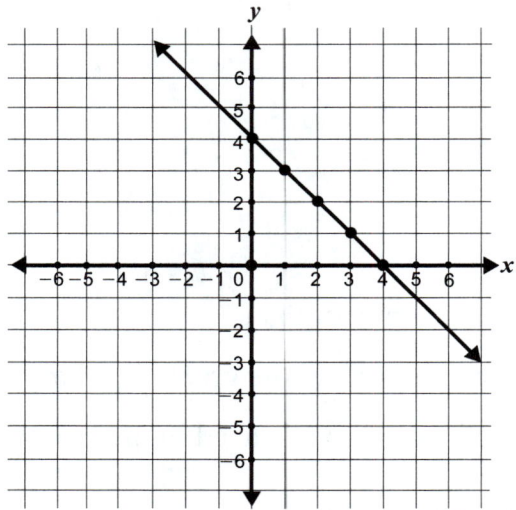

Some test questions will ask you to choose the correct graph of an equation. For others, you will be given a graph and asked to choose the correct equation.

Chapter 9: Algebra, Part 2

Let's Review 21: Linear Inequalities and Linear Equations

Complete each of the following questions. Use the Tip following each question to help you choose the correct answer. When you finish, check your answers with those at the end of Chapter 9.

1. The graph shown here is the graph of which of the following equations?

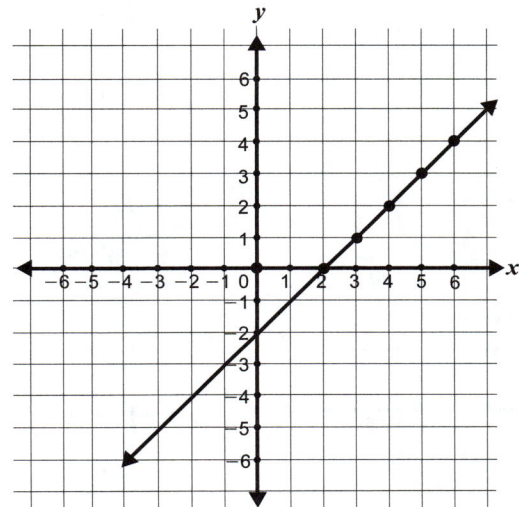

A. $y = x + 2$

B. $y = 2x - 1$

C. $y = 2x + 1$

D. $y = x - 2$

TIP: To find the correct equation, you need to list numbers that could fit into each equation and then plot them on the graph.

2. Which of these graphs is a graph of the equation $y - x = 1$?

A.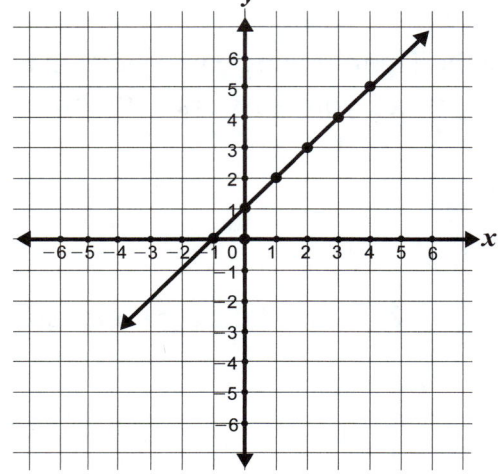

B.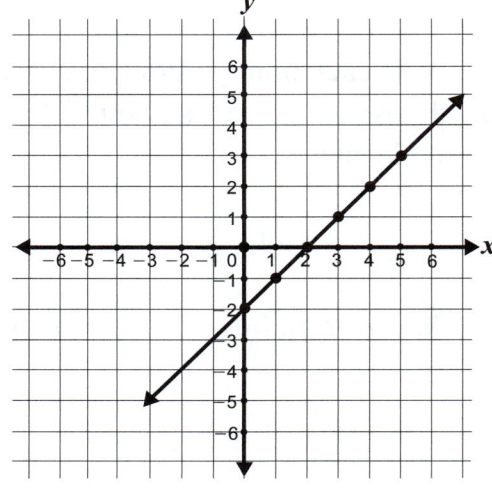

C.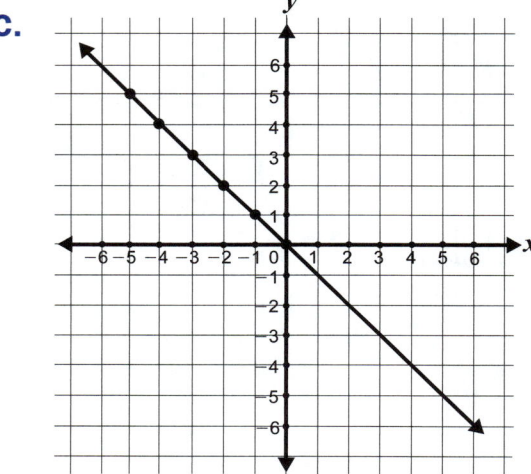

D.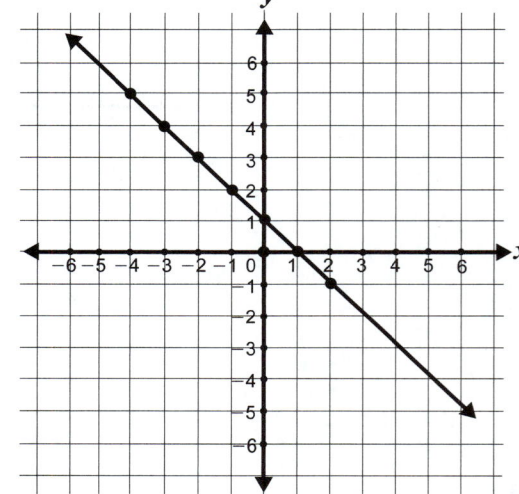

TIP: Substitute values for x and y in the equation. Then look at the line these numbers create on a graph.

Chapter 9: Algebra, Part 2

3. Given the inequality 3y < 18, solve for y.

   A. $y = 6$

   B. $y < 6$

   C. $y > 6$

   D. $y \leq 6$

TIP: The sign should be the same as the sign in the inequality.

Determining Slope

The **slope** of a line indicates a line's steepness; the greater the slope, the steeper the line. The slope of a line can be positive, negative, zero, or undefined. An **undefined slope** has a slope that cannot be determined. It might, for example, contain a zero, as in $\frac{3}{0}$.

Lines with a positive slope slant upward from left to right. Lines with a negative slope slant downward from left to right. And lines with 0 slope are horizontal lines. (They are not steep at all!) Horizontal lines have a 0 slope. Vertical lines have an undefined slope.

The **x-intercept** is the pair of coordinates at which a line crosses the x-axis, and the **y-intercept** is the pair of coordinates at which a line crosses the y-axis. You may be asked to find the x- and y-intercept on the GHSGT.

To determine the slope of a line, use the **rise over run formula:**

$$\frac{(y_2 - y_1)}{(x_2 - x_1)}$$

Suppose a line has the coordinates listed here. You would use the formula this way:

$$(1, 5) \text{ and } (4, -3)$$

$$(x_1, y_1), (x_2, y_2)$$

$$\frac{(-3 - 5)}{(4 - 1)}$$

The slope of this line is $-\frac{8}{3}$.

Let's Review 22: Determining Slope

Complete each of the following questions. Use the Tip following each question to help you choose the correct answer. When you finish, check your answers with those at the end of Chapter 9.

1. Which of the following describes the slope of a line parallel to the *x*-axis?

   A. positive slope

   B. negative slope

   C. zero slope

   D. undefined slope

   TIP If you don't remember what you learned about lines parallel to the *x*-axis, reread this section to find the answer.

2. What is the slope of a line that passes through the points (2, 5) and (6, 13)?

   A. -2

   B. 0

   C. 1

   D. 2

   TIP Use the rise-over-run formula to find the slope of this line.

3. Which of the following graphs describes a linear equation where the x-intercept is 0 and the y-intercept is 0?

A.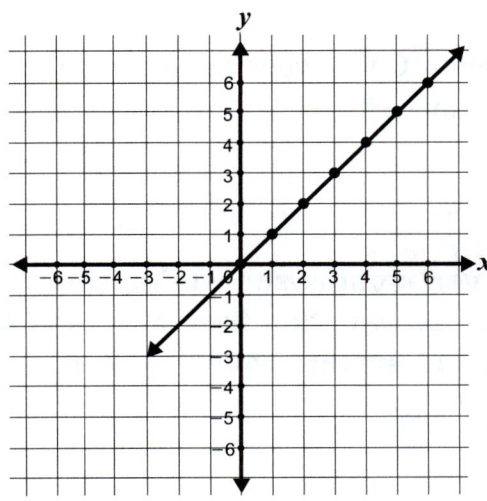

B.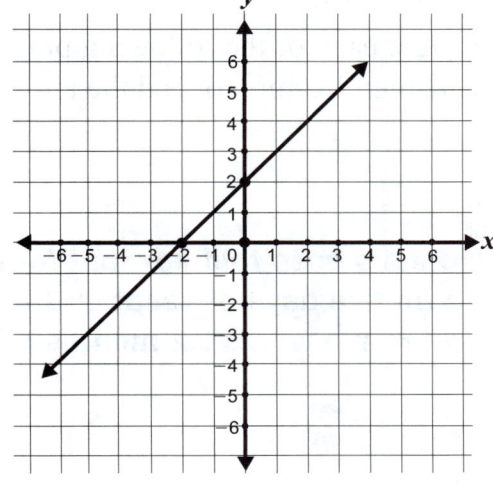

C.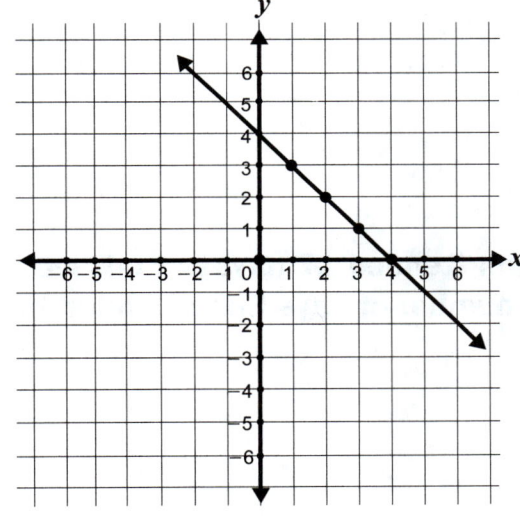

D.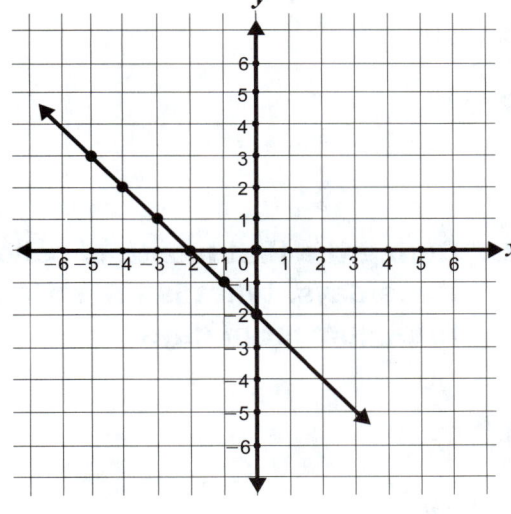

Chapter 9 Review

Complete each of the following practice problems. Check your answers at the end of this chapter. Be sure to read the answer explanations!

1. Chris's boss told her that for every seven months she works, she will earn five days of vacation time. How much vacation time will Chris earn after she works 12 months? Round to the nearest whole number.

   A. 8

   B. 9

   C. 10

   D. 12

2. During the first month of school, Erin attended 17 days but missed three days. What is the ratio of the number of days Erin attended to the total number of days?

   A. $\dfrac{3}{17}$

   B. $\dfrac{17}{3}$

   C. $\dfrac{3}{20}$

   D. $\dfrac{17}{20}$

3. Roberto drives 150 miles in five hours. Which proportion should Roberto use to determine how long it would take him to drive 500 miles?

A. $\dfrac{5}{500} = \dfrac{x}{150}$

B. $\dfrac{150}{5} = \dfrac{x}{500}$

C. $\dfrac{5}{150} = \dfrac{x}{500}$

D. $\dfrac{500}{x} = \dfrac{5}{150}$

4. This is the graph of which of the following equations?

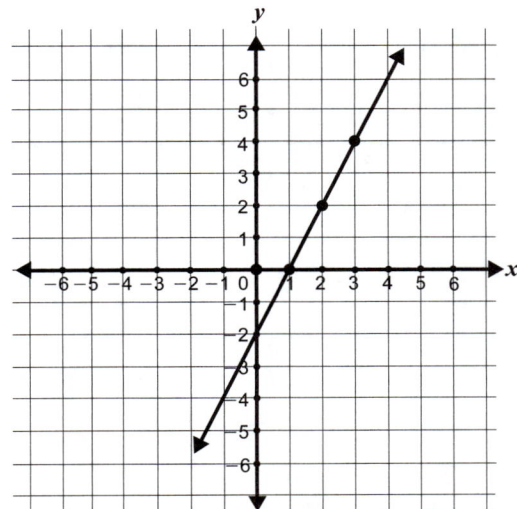

A. $y = 3x - 1$

B. $y = 2x - 2$

C. $y = 3x + 1$

D. $y = 2x + 2$

5. If 1 out of 10 people prefers hot dogs to pizza, how many people can be expected to prefer hot dogs in a town of 30,000 people?

A. 300

B. 3,000

C. 30,000

D. 300,000

6. Which of these graphs is a graph of the equation $x + y = 4$?

A.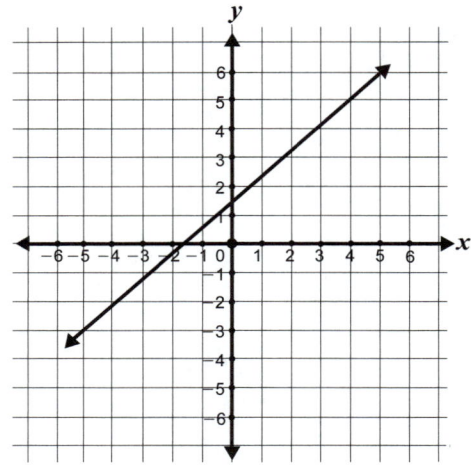

B.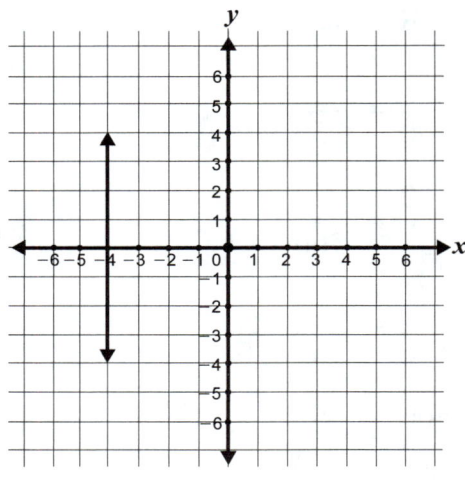

C.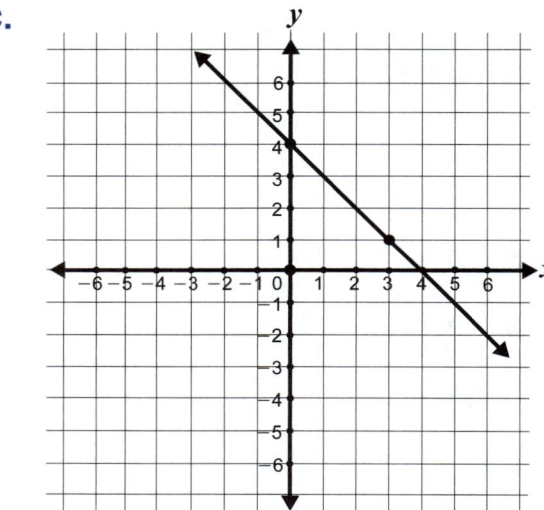

D.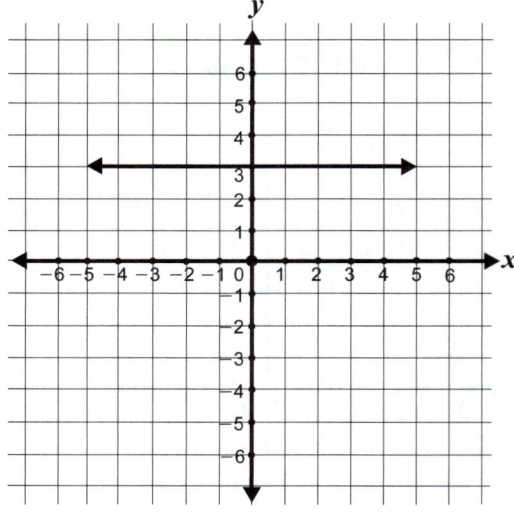

7. Which inequality describes the interval graphed on the number line?

A. $10 \geq g \geq 20$

B. $10 \leq g \leq 20$

C. $10 > g > 20$

D. $10 < g < 20$

Chapter 9: Algebra, Part 2

8. Which of the following describes the slope of a line parallel to the *y*-axis?

   A. positive slope

   B. negative slope

   C. zero slope

   D. undefined slope

9. Which is the slope of a line that passes through the points (2, 4) and (−7, 10)?

   A. $\dfrac{2}{3}$

   B. $-\dfrac{2}{3}$

   C. 2

   D. −2

Chapter 9 Answers

Let's Review 20: Proportions

1. D

If you put up the proportion like this: $\dfrac{4}{.80} = \dfrac{8}{x}$, cross multiply, and then put *x* on a side by itself, you'll see that the correct answer is D, $1.60.

2. C

The correct proportion is $\dfrac{3}{7} = \dfrac{S}{21}$.

3. B

The correct ratio has the number of b tablets as the numerator and the total number of tablets, $b + g$, as the denominator.

Let's Review 21: Linear Inequalities and Linear Equations

1. D

If you substitute some numbers for x and then subtract 2, you'll see that the answer choice D is correct.

2. A

Answer choice A is a graph of the equation $y - x = 1$. If you substitute some numbers for these variables, you'll see that the line in answer choice A is correct.

3. B

The solution should have the same sign as the inequality. Answer choice B is correct.

Let's Review 22: Determining Slope

1. C

Lines parallel to the x-axis have a slope of zero.

2. D

When you substitute the coordinates into the formula, you'll see that the slope is 2.

3. A

Both the x- and y-intercept of the graph shown in answer choice A are zero.

Chapter 9 Review

1. B

For this problem, the proportion should be set up this way: $\frac{5}{7} = \frac{x}{12}$.

Chapter 9: Algebra, Part 2

2. D

There were 20 days of school and Erin attended school on 17 of these days.

3. C

The proportion should be set up this way: $\frac{5}{150} = \frac{x}{500}$.

4. B

If you substitute values into the equation $y = 2x - 2$, you'll see that it matches the graph.

5. B

To solve this problem, divide 10 into 30,000.

6. C

Answer choice C is a graph of $x + y = 4$.

7. B

For this question, g is somewhere in the middle. Answer choice B—10 is less than or equal to g, which is less than or equal to 20—is the correct answer.

8. D

A line parallel to the y-axis is vertical. Vertical lines have an undefined slope.

9. B

If you substitute these coordinates into the formula to find the slope, you'll see that the slope is $-\frac{2}{3}$.

GHSGT Mathematics Practice Test

1

Practice Test 1 (GHSGT)

1. What is the volume of a cube that has an edge of 4 inches?

    A. 12 in.³
    B. 16 in.³
    C. 64 in.³
    D. 96 in.³

2. In a card game, Andy scored 22, 5, 22, 13, 12, 24, 24, 9, 20, and 19 points. What is the mean number of points Andy scored?

    A. 17
    B. 18
    C. 19
    D. 24

3. Which fraction is closest to .60?

    A. $\dfrac{1}{60}$

    B. $\dfrac{1}{4}$

    C. $\dfrac{1}{2}$

    D. $\dfrac{3}{4}$

4. If a bus travels at 65 miles per hour, how many miles will it travel in six hours?

   A. 71 miles
   B. 360 miles
   C. 390 miles
   D. 460 miles

5. On her first history test, Amy scored x points. On her second test, she scored 98 points, and on her third test she scored 80 points. Her total points for the three tests could be expressed as $x + (98 + 80)$.

   Use the associative property to write an equivalent expression.

   A. $(98 + 80) + x$
   B. $x = 98 + 80$
   C. $(x + 98) + 80$
   D. $x(98 + 80)$

6. An irregular hexagon has a perimeter of 36 inches. Five of its sides are 3, 4, 6, 8, and 12. What is the length of the remaining side?

   A. 3 inches
   B. 4 inches
   C. 6 inches
   D. 8 inches

7. The identity element for addition is 0. What is the identity element for multiplication?

   A. 0

   B. 1

   C. $\dfrac{1}{x}$

   D. $0.x$

8. **Number of Students in Emma's School**

Grade	Number of Students
1	20
2	28
3	27
4	22
5	25
6	26
7	27

Use the data in the chart to determine the median number of students.

- A. 20
- B. 22
- C. 25
- D. 26

9. You need to determine whether you have enough money for three items in your grocery cart. Which would be the most appropriate method to estimate the sum of the prices for these items?

- A. a calculator
- B. a computer
- C. mental arithmetic
- D. paper and pencil

10. Which value is the greatest?

- A. 3^4
- B. 2^5
- C. 5^3
- D. 6^2

11. Which answer shows the number that point B represents on the line?

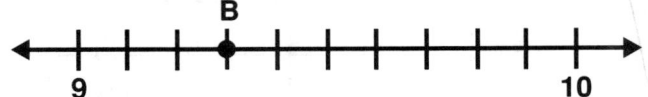

A. $9\dfrac{2}{9}$

B. $9\dfrac{2}{10}$

C. $9\dfrac{3}{9}$

D. $9\dfrac{3}{10}$

12. A school had 38 cheerleading uniforms at the beginning of the school year. If x represents the number of additional cheerleading uniforms purchased during the school year, which expression shows the number of cheerleading uniforms at the end of the school year?

A. $\dfrac{38}{x}$

B. $38 - x$

C. $38(x)$

D. $38 + x$

13. Give the coordinates of point P on the graph shown here.

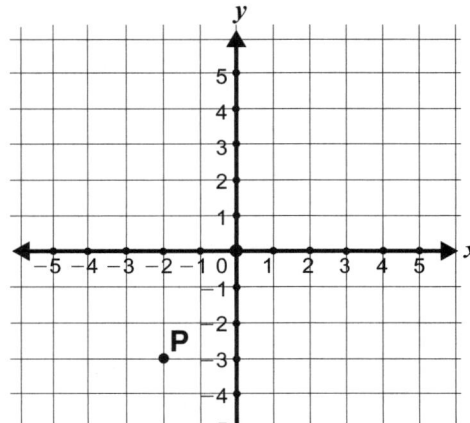

A. $(-3, -2)$
B. $(-2, -3)$
C. $(-2, -2)$
D. $(-3, -3)$

14. If the mean number of people who visited a museum over five days is 250, what is the total attendance during the five days?

    A. 750
    B. 1,000
    ✱ C. 1,250
    D. 2,500

15. Which of the following is the additive inverse of 80?

    A. $\dfrac{1}{80}$

    B. $-\dfrac{1}{80}$

    C. -80

    D. 1

16. Use the following tree diagram to predict the probability of flipping four coins and getting all heads or all tails.

    A. $\dfrac{1}{16}$

    B. $\dfrac{1}{8}$

    C. $\dfrac{1}{4}$

    D. $\dfrac{1}{2}$

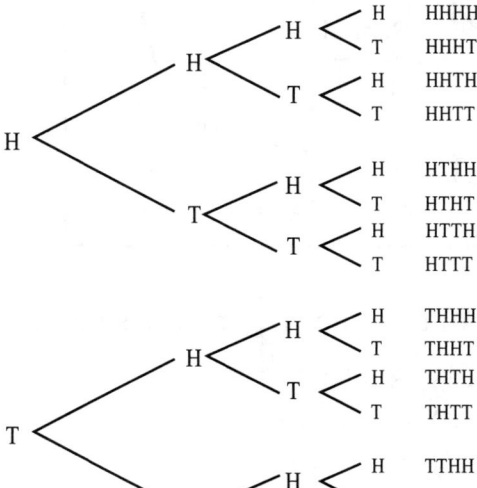

17. A block is **most** like a

    A. cone.
    B. cylinder.
    C. cube.
    D. sphere.

18. Which is the slope of a line that passes through the points (2, 2) and (−3, −3)?

    A. 1
    B. −1
    C. 2
    D. −2

19.

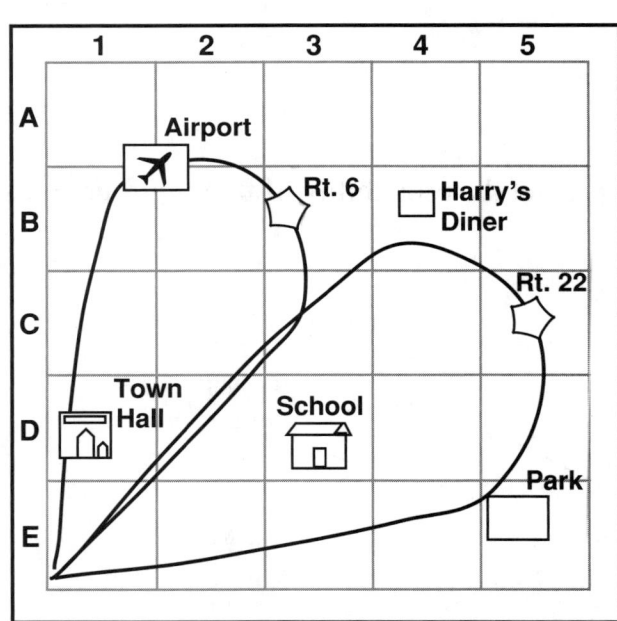

Which of the following indicates the square where Harry's Diner is located?

A. 3, B
B. 3, C
C. 4, A
D. 4, B

20. Find the missing length (*x*) of the pair of similar figures shown here.

    A. 8 feet
    B. 10 feet
    C. 12 feet
    D. 18 feet

    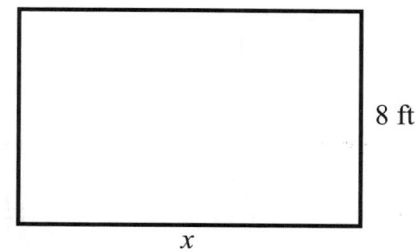

21. Simplify the following expression.

    $$x + 7x + 3y - y$$

    A. $8x + 2y$
    B. $10xy$
    C. $7x + 3y$
    D. $7x + 4y$

22. Which of the following numbers illustrates the inverse property of multiplication?

    A. $8 \times \dfrac{1}{8} = 1$

    B. $-8 + 8 = 0$

    C. $8 \times \dfrac{1}{8} = 0$

    D. $-8 \times 8 = 1$

23. Find the probability of spinning "blue" on the spinner.

    A. $\dfrac{1}{8}$

    B. $\dfrac{3}{8}$

    C. $\dfrac{1}{4}$

    D. $\dfrac{1}{2}$

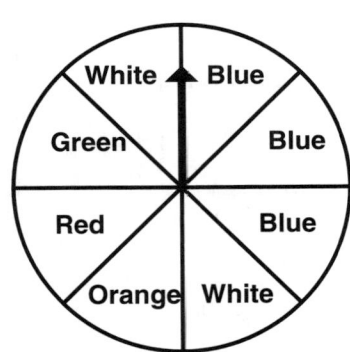

24. Which statement is represented by the algebraic expression $2 + x$?

    A. Daniel is x years old and his brother is two years younger.
    B. Keith is two years older than Bobby, who is x years old.
    C. Tracey is two times the age of her daughter, who is x years old.
    D. Jordan is two years younger than her sister Taylor, who is x years old.

25. The best estimate for the length of a person's leg is

    A. 3 inches.
    B. 3 feet.
    C. 3 yards.
    D. 3 miles.

26. Study Figures I and II. Which transformation, if any, of Figure I is shown in Figure II?

    A. no transformation
    B. reflection
    C. translation
    D. rotation

    P | q
    I   II

27. Find the numerical value of $2y + y^2$, when $y = 4$.

    A. 16
    B. 22
    C. 24
    D. 28

28. Estimate the difference: $21{,}567 - 10{,}204$

    A. 5,000
    B. 10,000
    C. 20,000
    D. 30,000

29. Kristen has a bag of 40 jelly beans. Five of these jelly beans are pink, 3 are blue, 10 are yellow, 2 are orange, 10 are green, 8 are black, and 2 are white. If Kristen reaches in without looking, what is the probability that she will pull out a white jelly bean?

    A. $\dfrac{1}{40}$

    B. $\dfrac{1}{20}$

    C. $\dfrac{1}{5}$

    D. $\dfrac{1}{4}$

30. If one out of five people exercises each day, how many people can be expected to exercise daily in a city of 25,000 people?

    A. 500
    B. 1,500
    C. 2,500
    D. 5,000

31. Give the coordinates of point A.

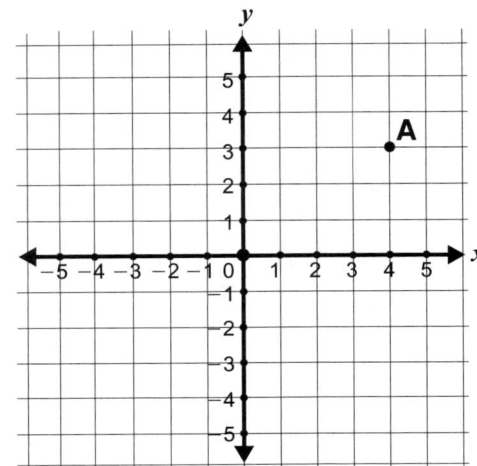

A. (4, 4)
B. (3, 3)
C. (3, 4)
D. (4, 3)

32. Dawn's English quiz score is shown by the expression $x + 10$, where x represents Tina's score. If Tina's score was 88, what is Dawn's score?

    A. 78
    B. 88
    C. 98
    D. 100

33. Which operation would be used to solve the equation $6y = 36$

    A. addition
    B. subtraction
    C. division
    D. multiplication

34. Emily works in a small crafts store where the cash register does not compute the sales tax. If the sales tax is 6%, what amount should Emily add to a purchase of $12.00?

    A. $0.60
    B. $0.72
    C. $6.00
    D. $7.20

35. There are 20 straws in a box; some are red and some are blue. The probability of reaching into the box and pulling out a red straw is $\frac{3}{5}$. How many blue straws are in the box?

   A. 3
   B. 6
   C. 8
   D. 12

36. Which of the following algebraic expressions corresponds to "*a* decreased by 4 is equal to the product of 2 and 4"?

   A. $a - 4 = 2 + 4$

   B. $a - 4 = \frac{2}{4}$

   C. $\frac{a}{4} = 2 \times 4$

   D. $a - 4 = 2 \times 4$

37. Which point on the graph has the coordinates $(4, -2)$?

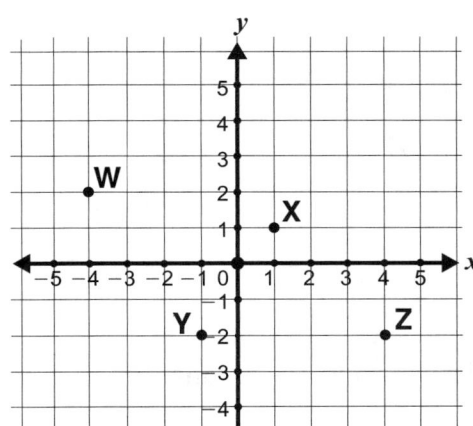

   A. point W
   B. point X
   C. point Y
   D. point Z

38. Solve for x, if $2x = 16$.

   A. 6
   B. 7
   C. 8
   D. 9

39. Richard recorded the number of cars that drove past his house on each of ten days. He recorded his findings in the following chart.

Day	Number of Cars
1	16
2	3
3	3
4	3
5	8
6	24
7	16
8	9
9	11
10	11

   What is the range in the number of cars?

   A. 3
   B. 11
   C. 13
   D. 21

40. Moving a geometric figure so that it is flipped across a line is a transformation by

   A. inversion.
   B. reflection.
   C. rotation.
   D. translation.

41. A 120-pound container of grain sells for $30. How much would a 150-pound container of grain cost?

   A. $32.50
   B. $35.00
   C. $37.50
   D. $40.00

42. The number of students enrolled in introductory courses at a university is shown on the following graph. How many more students are enrolled in the composition course than in psychology?

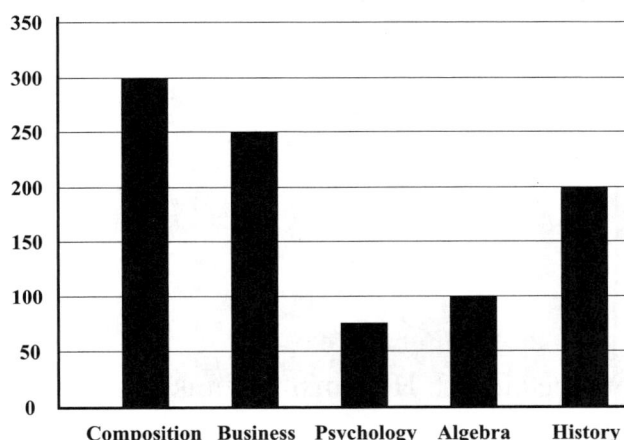

College Enrollment in Introductory Courses at Anytown University

   A. 200
   B. 225
   C. 250
   D. 375

43. A business seminar begins at 10:30 A.M. and ends at 3:00 P.M. If people attending the seminar take 30 minutes for lunch, how long is this seminar?

   A. 3 hours

   B. $3\frac{1}{2}$ hours

   C. 4 hours

   D. $4\frac{1}{2}$ hours

44. What is the approximate area of a circle with a radius of 8 cm?
   (Use $A = \pi r^2$, and $\pi = 3.14$.)

   A. 25 cm²
   B. 50 cm²
   C. 201 cm²
   D. 402 cm²

45. Jeremy lives six miles from the airport. How many feet is this?

   A. 5,286 feet
   B. 15,840 feet
   C. 26,400 feet
   D. 31,680 feet

46. What is the value of x, if $2x + 6 = 30$?

   A. 10
   B. 12
   C. 14
   D. 22

47. A storage tank holds 100 liters of liquid. How many hectoliters is this?

   A. $\frac{1}{2}$

   B. 1

   C. 10

   D. 100

48. Solve for x:

   $9x - 5 = 5 \times 5 + 15$

   A. 4
   B. 5
   C. 7
   D. 40

49. A rotating sprinkler is used to water a yard. The radius of the area being sprayed is 12 feet. What is the approximate wet area of the yard?

   (Use $A = \pi r^2$, and $\pi = 3.14$.)

   A. 38 feet^2
   B. 120 feet^2
   C. 144 feet^2
   D. 452 feet^2

50. Which of the following algebraic expressions corresponds to "*x* increased by 4 is equal to one-fourth of *y*"?

   A. $4x = \dfrac{y}{4}$

   B. $x - 4 = \dfrac{y}{4}$

   C. $x + 4 = \dfrac{y}{4}$

   D. $x + 4 = 4y$

51. Find the numerical value of $4y^2 + 9$, when $y = 2$.

   A. 17
   B. 20
   C. 24
   D. 25

52. To determine the weight of an elephant, which is the most appropriate unit of measure?

   A. gallons
   B. ounces
   C. liters
   D. tons

53. If a penny is tossed five times and on the first two tosses it comes up heads, what is the probability of getting heads on the third toss?

   A. $\dfrac{1}{5}$

   B. $\dfrac{1}{4}$

C. $\dfrac{1}{3}$

D. $\dfrac{1}{2}$

54. Santo rides a bus from his school to a basketball tournament in three hours. If he knows that bus's average speed is 45 mph and he wants to find the distance from his school to the basketball tournament, what equation should he use?

(Use the formula $d = rt$.)

A. $d = 3 + 45$

B. $d = \dfrac{45}{3}$

C. $3d = 45$

D. $d = 45 \times 3$

55. The results of a survey asking how students get to school is shown.

Bus	58%
Drive	22%
Walk	10%
Other	10%

Which type of graph should be used to show the results of the survey?

A. bar graph
B. circle graph
C. pictograph
D. line graph

56. A high-speed train travels 500 kilometers an hour. How far will it travel in $3\frac{1}{2}$ hours?

    A. 1,500 kilometers
    B. 1,625 kilometers
    C. 1,750 kilometers
    D. 2,000 kilometers

57. Pilar spent a total of $164 for six sweaters. Later she bought another sweater. She spent an average of $28.00 per sweater for the seven sweaters. What did she pay for the seventh sweater?

    A. $28.00
    B. $30.00
    C. $32.00
    D. $36.00

58. Deanne is making a wallpaper border using the triangle shown here. Classify the triangle according to the lengths of its sides.

    A. acute triangle
    B. equilateral triangle
    C. isosceles triangle
    D. right triangle

59. The lines in the letter × are examples of what kind of line segments?

    A. parallel
    B. collinear
    C. angles
    D. perpendicular

60. A square piece of paper, each of whose sides is six inches long, is folded diagonally on the dotted line, as shown. To the nearest inch, how long is the crease made in the fold?

    A. 6 inches
    B. 8 inches
    C. 36 inches
    D. 72 inches

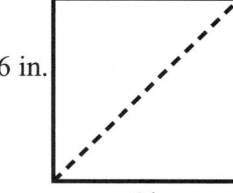

Answer Explanations

1. C (Standard Assessed: Measurement and Geometry)
A cube has the same length, width, and height, so you would substitute 4 in. into the formula $V = lwh$. $V = (4 \times 4)(4) = 64$ in.^{3}

2. A (Standard Assessed: Data Analysis)
To find the mean number of points Andy scored, total his points (170). Then divide by 10. The answer is 17.

3. C (Standard Assessed: Numbers and Computation)
To convert a decimal to a fraction, move the decimal point two places to the right and put the number over 100 and reduce. Therefore, $.60 = \frac{60}{100}$, which is closest to $\frac{1}{2}$.

4. C (Standard Assessed: Data Analysis)
To find out how far a bus traveling at 65 mph will travel in six hours, multiply the numbers. The answer is 390.

5. C (Standard Assessed: Numbers and Computation)
According to the associative property, parentheses can be moved. Therefore, $x + (98 + 80)$ is the same as $(x + 98) + 80$.

6. A (Standard Assessed: Measurement and Geometry)
If the perimeter of the hexagon is 36, add 3, 4, 6, 8, and 12 and subtract this amount from 36. This is the length of the remaining side.

7. B (Standard Assessed: Numbers and Computation)
The identity element for multiplication is 1.

8. D (Standard Assessed: Data Analysis)
To find the median number of students, put the number of students in each grade in order from least to greatest. The number in the middle, in this case 26, is the median.

9. C (Standard Assessed: Numbers and Computation)
If there are only three items in your grocery cart, you should be able to use mental arithmetic to see whether you have enough money.

10. C   (Standard Assessed: Numbers and Computation)
You should be able to quickly determine which number is greatest by using a calculator. Answer choice A is 81, answer choice B is 32, answer choice C is 125, and answer choice D is 36.

11. D   (Standard Assessed: Numbers and Computation)
There are ten marks between numbers 9 and 10 on the number line. Point B is at the third mark. So the fraction it represents is $9\frac{3}{10}$.

12. D   (Standard Assessed: Algebra)
If the school began the year with 38 cheerleading uniforms and then purchased more (x), the correct expression is $38 + x$.

13. B   (Standard Assessed: Measurement and Geometry)
The coordinates of point P are $(-2, -3)$.

14. C   (Standard Assessed: Data Analysis)
To find out how many people attended the museum over five days, multiply the mean by 5.

15. C   (Standard Assessed: Numbers and Computation)
The additive inverse is the opposite. The opposite of 80 is -80.

16. B   (Standard Assessed: Data Analysis)
There are two instances of all tails (TTTT) or all heads (HHHH) out of 16 possibilities, so the correct answer is $\frac{2}{16}$, which reduces to $\frac{1}{8}$.

17. C   (Standard Assessed: Measurement and Geometry)
A block is a cube.

18. A   (Standard Assessed: Algebra)
To find the slope of the line, use this formula: $\frac{y_2 - y_1}{x_2 - x_1}$ or $\frac{2-(-3)}{2-(-3)}$. The answer is $\frac{5}{5}$ or 1.

19. D   (Standard Assessed: Measurement and Geometry)
Harry's Diner is located at 4, B.

20. C (Standard Assessed: Measurement and Geometry)
The measurements of the second rectangle are double the first, so the length of the second rectangle is 12 feet.

21. A (Standard Assessed: Algebra)
To simplify $x + 7x + 3y - y$, combine like terms: $8x + 2y$.

22. A (Standard Assessed: Numbers and Computation)
The inverse property of multiplication involves the number 1. You need to choose the answer choice in which the result of multiplication is 1: $\frac{8}{1} \times \frac{1}{8} = \frac{8}{8}$ or 1.

23. B (Standard Assessed: Data Analysis)
There are eight sections of the spinner, so this is the denominator. Three sections are blue, so the probability of spinning blue is $\frac{3}{8}$.

24. B (Standard Assessed: Algebra)
The only situation that is the same as $2 + x$ is answer choice B: Keith is two years older than Bobby, who is x years old.

25. B (Standard Assessed: Numbers and Computation)
A person's leg is about 3 feet long.

26. B (Standard Assessed: Measurement and Geometry)
The letter P is reflected across the line; the reflection is a mirror image.

27. C (Standard Assessed: Algebra)
If you substitute 4 into the expression: $2y + y^2$, it looks like this: $2(4) + 4^2$ or $8 + 16 = 24$.

28. B (Standard Assessed: Numbers and Computation)
To estimate the difference between these numbers, round them to the nearest ten thousand: $20,000 - 10,000 = 10,000$.

29. B (Standard Assessed: Data Analysis)
There are 40 jelly beans in the bag, so this is the denominator. Only two of them are white, so 2 is the numerator. When you reduce the fraction, you get $\frac{1}{20}$.

30. D (Standard Assessed: Algebra)
If one out of five people exercises and you want to determine how many people out of 25,000 exercise, divide 5 into 25,000. The answer is 5,000.

31. D (Standard Assessed: Measurement and Geometry)
The coordinates of point A are (4, 3).

32. C (Standard Assessed: Algebra)
Dawn's score is ten points higher than Tina's. If Tina scored 88, Dawn scored 98.

33. C (Standard Assessed: Algebra)
To solve the equation $6y = 36$, divide both sides by 6: $y = \frac{36}{6} = 6$.

34. B (Standard Assessed: Numbers and Computation)
To find the amount of tax Emily needs to add to the purchase, multiply $12 by .06. The answer is $0.72.

35. C (Standard Assessed: Data Analysis)
Since there are 20 straws in the box, convert $\frac{3}{5}$, the number of red straws, into a fraction with 20 as the denominator: $\frac{12}{20}$. Eight straws are blue.

36. D (Standard Assessed: Algebra)
The phrase "*a* decreased by 4 is equal to the product of 2 and 4" is the same as $a - 4 = 2 \times 4$.

37. D (Standard Assessed: Measurement and Geometry)
Point Z is located at (4, −2).

38. C (Standard Assessed: Algebra)
To solve this problem, divide both sides of the equation by 2: $x = \frac{16}{2}$ or $x = 8$.

39. D (Standard Assessed: Data Analysis)
To find the range, subtract the fewest number of cars, 3, from the greatest number of cars, 24. The answer is 21.

40. B (Standard Assessed: Measurement and Geometry)
Moving a figure so that it is flipped across a line is a reflection.

41. C (Standard Assessed: Algebra)
Set up a proportion to find this answer: $\frac{120}{30} = \frac{150}{x}$. Then cross multiply: $150 \times 30 = 4{,}500 = 120x$. Then divide 120 into 4,500. $x = \$37.50$.

42. B (Standard Assessed: Data Analysis)
About 300 students are enrolled in composition and 75 are enrolled in psychology. If you subtract 75 from 300, you get 225.

43. C (Standard Assessed: Measurement and Geometry)
The time between 10:30 A.M. and 3:00 P.M. is $4\frac{1}{2}$ hours. If you deduct 30 minutes, the amount of time for lunch, you find that the length of the seminar is four hours.

44. C (Standard Assessed: Algebra)
The radius of the circle is 8. Substitute this into the formula πr^2. The number 8 squared is 64, and 64×3.14 is about 201.

45. D (Standard Assessed: Measurement and Geometry)
There are 5,280 feet in a mile. If you multiply this number by 6, you get 31,680.

46. B (Standard Assessed: Algebra)
To solve this equation, put x on one side of the equation: $x = \frac{(30 - 6)}{2}$. The answer is 12.

47. B   (Standard Assessed: Measurement and Geometry)
There are 100 liters in a hectoliter, so the answer is one.

48. B   (Standard Assessed: Algebra)
Simplify the equation before solving for x. $9x - 5 = 40$. Now put x on one side: $x = \frac{(40 + 5)}{9}$. The answer is 5.

49. D   (Standard Assessed: Measurement and Geometry)
To solve this problem, you need to find the area of a circle. To do this, use the formula $A = \pi r^2$. The radius is 12, and 12 squared is 144. The number 144 multiplied by 3.14 is about 452.

50. C   (Standard Assessed: Algebra)
The expression "x increased by 4 is equal to one-fourth of y" is the same as $x + 4 = \frac{y}{4}$.

51. D   (Standard Assessed: Algebra)
To solve this problem, substitute 2 for y: $4(2)^2 + 9$. The answer is 25.

52. D   (Standard Assessed: Measurement and Geometry)
An elephant weighs several tons.

53. D   (Standard Assessed: Data Analysis)
The odds of tossing either heads or tails is always $\frac{1}{2}$.

54. D   (Standard Assessed: Algebra)
To solve this problem, insert the rate, 45, and the time, 3, into the formula $d = rt$. Then the equation is $d = 45 \times 3$.

55. B   (Standard Assessed: Data Analysis)
A circle graph is the best way to show parts of a whole.

56. C   (Standard Assessed: Measurement and Geometry)
To solve this problem, multiply 500 by 3.5. The answer is 1,750.

57. C   (Standard Assessed: Data Analysis)
To find the answer to this problem, multiply the average cost for the seven sweaters, $28.00 by 7. Then subtract $164 from this amount.

58. D   (Standard Assessed: Measurement and Geometry)
This triangle has a 90-degree angle, so it is a right triangle.

59. D   (Standard Assessed: Measurement and Geometry)
The lines in the letter × are perpendicular. They intersect to form four right angles.

60. B   (Standard Assessed: Measurement and Geometry)
You have to use the Pythagorean theorem ($a^2 + b^2 = c^2$) to solve this problem. Both side a and side b of the triangle measure 6 inches, so $36 + 36 = 72$. Then you have to find the square root of 72 to get the correct answer: about 8 inches.

GHSGT Mathematics Practice Test

2

Practice Test 2

1. Frankie had 30 baseball cards at the beginning of the week. If x represents the number of baseball cards Frankie gave to his friend Amy on Tuesday and y represents the number of baseball cards his mother gave him on Thursday, which expression shows the number of baseball cards Frankie has at the end of the week?

    A. $x + 30 - y$
    B. $30 - x + y$
    C. $30x - y$
    D. $30y + x$

2. Find the probability of spinning a 2 on the spinner.

    A. $\dfrac{1}{6}$

    B. $\dfrac{1}{4}$

    C. $\dfrac{1}{3}$

    D. $\dfrac{1}{2}$

    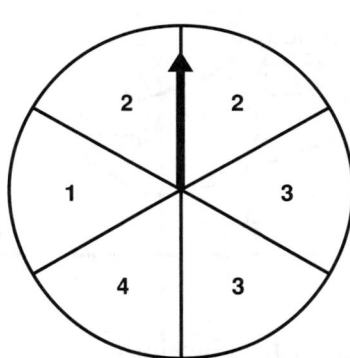

3. Which of the following algebraic expressions corresponds to "four added to the product of a number and 6"?

    A. $4(6 + x)$

    B. $\dfrac{6x}{4}$

    C. $4 + 6x$

    D. $10x$

4. A good estimate for the length of Melissa's desk would be

   A. 4 inches.
   B. 4 feet.
   C. 4 yards.
   D. 4 miles.

5.

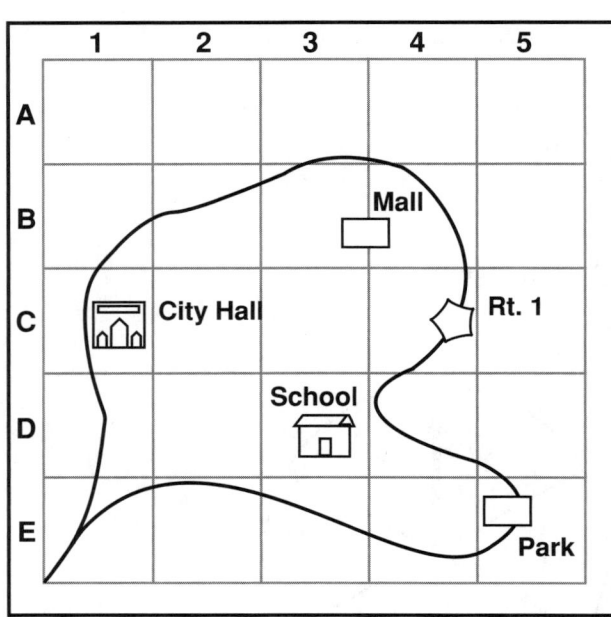

Which of the following indicates the square where City Hall is located?

   A. 1, B
   B. 1, C
   C. 2, B
   D. 2, C

6. What is another way to express 225?

   A. 5^4
   B. 15^2
   C. 22.5×10^2
   D. 12^3

7. A work shift starts at 6:30 A.M. and stops at 2:30 P.M. Workers take 60 minutes for lunch. What is the length of their workday?

   A. 7 hours
   B. 7.5 hours
   C. 8 hours
   D. 8.5 hours

8. Given the inequality $6y < 42$, solve for y.

   A. $y = 7$
   B. $y < 7$
   C. $y > 7$
   D. $y \leq 7$

9. Lisa has a bag of 30 marbles. Five of these marbles are white, three are blue, ten are pink, five are red, two are green, three are orange, and two are black. If Lisa reaches into the bag and pulls out a marble without looking, what is the probability that she will pull out a red marble?

   A. $\dfrac{1}{30}$
   B. $\dfrac{1}{6}$
   C. $\dfrac{1}{5}$
   D. $\dfrac{1}{4}$

10. Simplify the following expression:

    $\dfrac{3y}{y^3}$ ($y \neq 0$)

    A. $3y^2$

    B. $\dfrac{3y}{y}$

    C. $\dfrac{3}{y^2}$

    D. $\dfrac{3}{y3}$

11. There are 20 coins in a box; some are pennies and some are nickels. The probability of reaching into the box and selecting a nickel is $\dfrac{3}{10}$. How many pennies are in the box?

    A. 5
    B. 10
    C. 14
    D. 16

12. Which point on the following graph has the coordinates $(-1, 2)$?

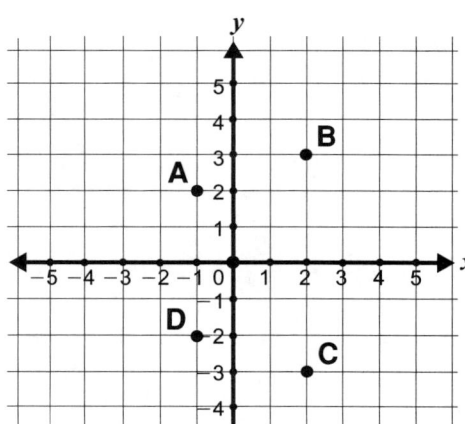

    A. point A
    B. point B
    C. point C
    D. point D

13. Given that water boils at 212°F and freezes at 32°F, what would be the most likely temperature of the water in a swimming pool during the summer?

    A. 35°F
    B. 50°F
    C. 85°F
    D. 170°F

14. What is .75 expressed as a percent?

    A. .75%
    B. 7.5%
    C. 75%
    D. 750%

15. Karen's house is 2,000 meters from her school. How many kilometers is this distance?

    A. $1\frac{1}{2}$

    B. 2

    C. 20

    D. 200

16. Lindsay's new car uses six gallons of gasoline to drive 150 miles. Which proportion should Lindsay use to determine the number of gallons of gasoline (G) she'll need to drive 200 miles?

   A. $\dfrac{6}{G} = \dfrac{200}{150}$

   B. $\dfrac{6}{150} = \dfrac{G}{200}$

   C. $\dfrac{G}{150} = \dfrac{6}{200}$

   D. $\dfrac{6}{150} = \dfrac{200}{G}$

17. What is the volume of the box pictured here?

   A. 22 in.³
   B. 72 in.³
   C. 288 in.³
   D. 576 in.³

18. Tito is going to roll a six-sided number cube. What is the probability of rolling an odd number?

   A. $\dfrac{1}{6}$

   B. $\dfrac{1}{4}$

   C. $\dfrac{1}{3}$

   D. $\dfrac{1}{2}$

19. Melanie uses the expression $8a + 12b$ to determine the amount she earns at a pay rate of $8 an hour, plus time and a half for overtime. One week she worked 40 hours, plus 8 hours of overtime. What is her total pay for the week?

   A. $68
   B. $120
   C. $320
   D. $416

20. Give the coordinates of point D on the following graph.

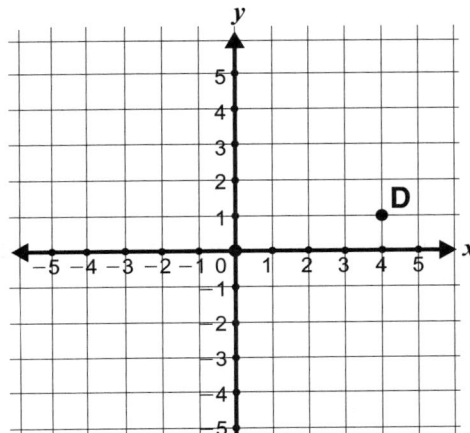

   A. (1, 4)
   B. (4, 1)
   C. (0, 4)
   D. (4, 0)

21. Sarah plans to give each student in her class three stickers. There are 24 students in her class. The number of stickers she needs can be expressed as 3×24.

   Use the commutative property to write an equivalent expression.

   A. $3 + 24$

   B. $\dfrac{3}{24}$

   C. $3 \times .24$

   D. 24×3

22. Find the numerical value of $2y + 10x$, when $y = 4$ and $x = 6$.

   A. 22
   B. 40
   C. 60
   D. 68

23. If a commercial jet travels 570 miles per hour, how many miles will it travel in four hours?

   A. 143 miles
   B. 1,140 miles
   C. 2,280 miles
   D. 2,850 miles

24. Ms. Roberts constructed a diagram to illustrate the number of girls who are enrolled in the athletic program.

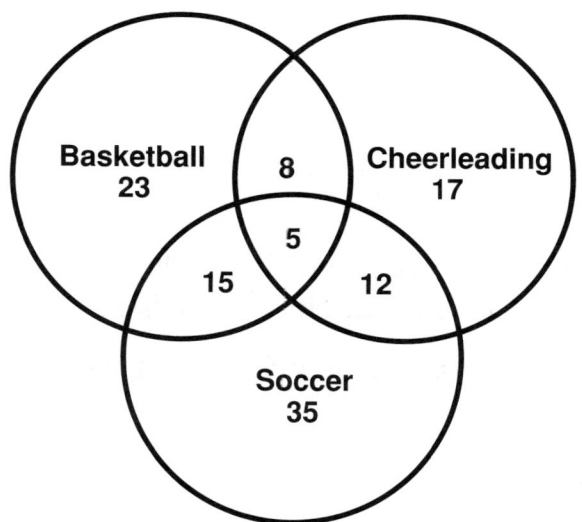

How many girls play both basketball and soccer but are not involved with cheerleading?

   A. 8
   B. 12
   C. 15
   D. 35

25. How many girls play both basketball and soccer and are on the cheerleading squad?

    A. 5
    B. 8
    C. 17
    D. 75

26. Alex is on the cross-country team. She ran in five meets and did not run in two. What is the ratio of the number of meets she ran in to the total number of meets?

    A. $\dfrac{2}{5}$

    B. $\dfrac{7}{5}$

    C. $3\dfrac{3}{5}$

    D. $\dfrac{5}{7}$

27. The identity element for multiplication is 1. What is the identity element for addition?

    A. 0

    B. 1

    C. $\dfrac{1}{x}$

    D. $0 + x$

28. Find the numerical value of $2y^3 + 4$, when $y = 2$.

    A. 12
    B. 14
    C. 20
    D. 24

29. Beth scored 18, 18, 15, 18, 18, 24, 21, 20, 24, and 14 points during her first ten basketball games. What is her mean score?

    A. 10
    B. 18
    C. 19
    D. 21

30. Which of the following is the additive inverse of -10?

    A. -10

    B. $\dfrac{1}{10}$

    C. 1

    D. 10

31. Jose's scores in psychology class are 94, 69, 84, 78, 90, 75, 94, 90, 90, and 95. What is the mode of his test scores?

    A. 94
    B. 88
    C. 90
    D. 92

32. The number of blue marbles Renee has is shown by the expression $2x - 3$, with x representing the number of white marbles. If Renee has 14 white marbles, how many blue marbles does she have?

   A. 13
   B. 14
   C. 25
   D. 28

33. To calculate a company's taxes, which method is most appropriate?

   A. paper and pencil
   B. calculator
   C. mental arithmetic
   D. computer

34. The perimeter of a triangle is 28. Two of its sides measure 10 and 8. What is the length of the remaining side?

   A. 6
   B. 8
   C. 10
   D. 12

35. Find the numerical value of $2y^3 - 5$, when $y = 3$.

   A. 22
   B. 40
   C. 49
   D. 54

36. Which operation would be used to solve the equation $y + 9 = 12$?

   A. addition
   B. subtraction
   C. division
   D. multiplication

37. What is another way to express 250,000?

   A. 25^4
   B. 50^3
   C. 2.5×10^4
   D. 2.5×10^5

38. Which statement is represented by the algebraic expression $12 - y$?

   A. Tyler has y amount of stamps but then gives 12 away. How many stamps does Tyler have now?
   B. Kate has a stamp collection and wants to give 12 stamps to y number of students in her class. How many students will get stamps?
   C. Rachel has 12 stamps and her mother gives her y more. How many stamps does Rachel have now?
   D. Hakim has 12 stamps but gives y stamps to his brother. How many stamps does Hakim have left?

39. Choose the situation in which a result using approximate numbers would be expected.

   A. a student's grade point average in a subject for a year
   B. the number of students in a school who play a sport
   C. the cost of a television cable bill for one month
   D. the number of votes a politician receives in an election

40. Which type of graph is best used to show the number of people in five towns who receive a morning newspaper?

   A. bar graph
   B. circle graph
   C. line graph
   D. Venn diagram

41. Cecily has a photograph that measures 8 inches wide and 10 inches in length. If Cecily has the photograph enlarged so that it is 24 inches wide, how long will the picture be?

    A. 24 inches
    B. 30 inches
    C. 240 inches
    D. 320 inches

42. Find the numerical value of $7x^2 + 10$, when $x = 4$.

    A. 38
    B. 66
    C. 112
    D. 122

43. To determine the weight of a small piece of cloth, which is the most appropriate unit of measure?

    A. ounces
    B. pounds
    C. gallons
    D. tons

44. Estimate the sum of 41, 12, 19, 18, and 24.

    A. 90
    B. 100
    C. 110
    D. 120

45. The oldest person in an audience of 100 is 62. If the range is 51, what is the age of the youngest member of the audience?

    A. 10
    B. 11
    C. 12
    D. 13

46. Salma's age is shown by the expression $y - 3$, where y represents Malcolm's age. If Malcolm is 17, how old is Salma?

    A. 13
    B. 14
    C. 21
    D. 51

47. In which figure could the Pythagorean theorem be used to find the length of $\overline{YZ}$?

    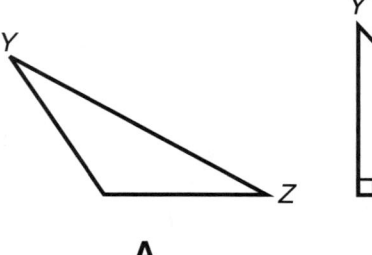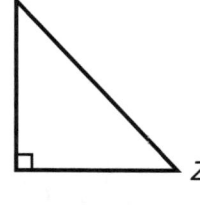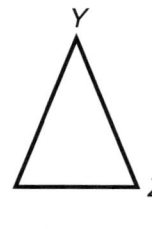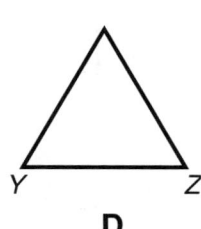

    A   B   C   D

    A. A
    B. B
    C. C
    D. D

48. Madeline earns $8 an hour babysitting her cousins during the 10 weeks of summer vacation. If she averages 12 hours per week, what is a reasonable estimate of what Madeline will earn during the summer?

    A. $100
    B. $500
    C. $800
    D. $1,000

49. Give the coordinates of point A on the following graph.

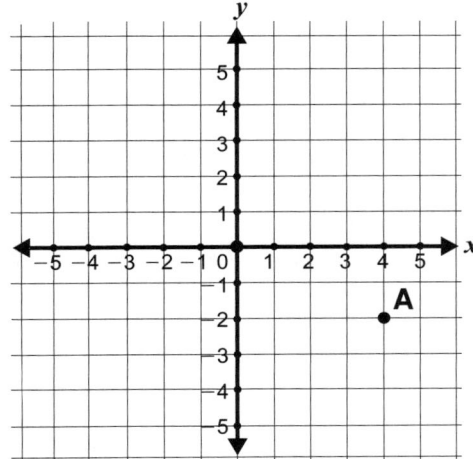

A. (4, −2)
B. (5, −2)
C. (−2, 4)
D. (−2, 5)

50. Study Figures I and II. Which transformation, if any, of Figure I is shown in Figure II?

A. no transformation
B. reflection
C. translation
D. rotation

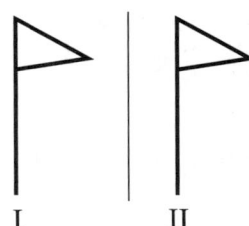

51. The mass of a box of cereal can be best measured in

A. grams.
B. hectograms.
C. kilograms.
D. milligrams.

52. Which of the following algebraic expressions corresponds to "x decreased by 2 is equal to $\frac{1}{3}$ of y"?

 A. $x - 2 = \frac{y}{3}$

 B. $x - 2 = 3y$

 C. $x = 2 - \frac{y}{3}$

 D. $-2x = -\frac{y}{3}$

53. Determine the approximate volume of a cylinder with a radius of 3 in. and a height of 12 in. Use 3.14 for π.

 $V = \pi r^2 h$

 A. 108 in.³
 B. 113 in.³
 C. 226 in.³
 D. 339 in.³

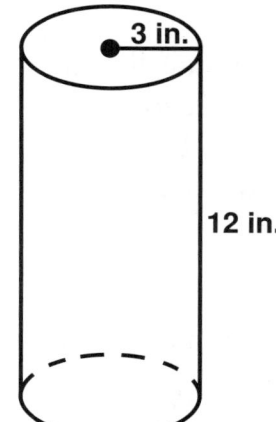

54. If $\angle ACB$ measures 45°, what is the measure of $\angle BCD$?

 A. 105°
 B. 115°
 C. 125°
 D. 135°

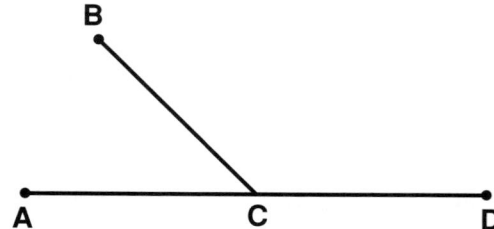

55. The number of students in Bibi's class is shown by the expression $3x - 2$, with x representing the number of girls. If the class has 10 girls, how many boys does it have?

    A. 32
    B. 28
    C. 18
    D. 12

56. Which point on the following graph has the coordinates $(-1, 1)$?

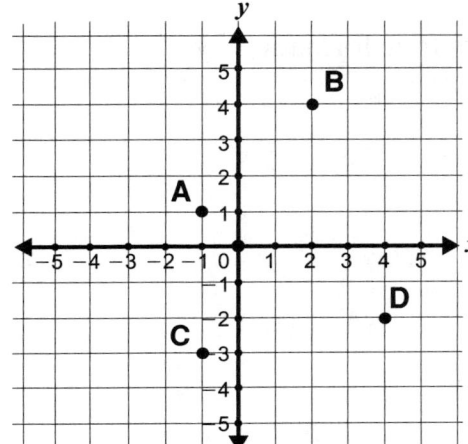

    A. point A
    B. point B
    C. point C
    D. point D

57. If a jacket originally cost $75 and is selling at a 20% discount, what is the amount of the discount?

    A. $7.50
    B. $11.25
    C. $15.00
    D. $18.75

58. What is the value of y, if $2y - 4 = 16$?

    A. 4
    B. 6
    C. 8
    D. 10

59. The volume of a cylinder is found using the formula: $V = \pi r^2 h$.

What is the approximate volume of the cylinder shown here?

A. 100 in³
B. 128 in³
C. 402 in³
D. 804 in³

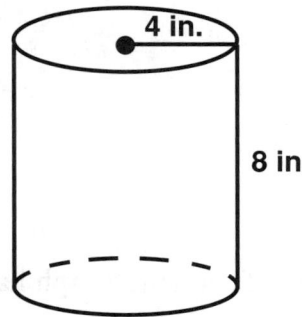

60. Moving a geometric figure in one direction is transformation by

A. translation.
B. inversion.
C. rotation.
D. reflection.

Answer Explanations

1. B (Standard Assessed: Algebra)
Frankie had 30 baseball cards at the beginning of the week and he gave x away. So, you know the expression will begin with $30 - x$. Then Frankie's mother gives him more baseball cards (y). The entire expression should be $30 - x + y$.

2. C (Standard Assessed: Data Analysis)
There are six sections on the spinner, so the denominator will be 6. Two of these sections have a 2 on them, so the fraction is $\frac{2}{6}$ or $\frac{1}{3}$.

3. C (Standard Assessed: Algebra)
The phrase "4 added to the product of a number and 6" is $4 + 6x$, since the product is a result of multiplication.

4. B (Standard Assessed: Measurement and Geometry)
A desk is about 4 feet long. If you were unsure, you could find this answer by process of elimination. Four inches (answer choice A) is too short and the other answer choices are too long.

5. B (Standard Assessed: Measurement and Geometry)
City Hall is located in block 1, C.

6. B (Standard Assessed: Numbers and Computation)
You can use your calculator to quickly determine this answer: 225 is 15^2.

7. A (Standard Assessed: Measurement and Geometry)
The time of the work shift is eight hours minus one hour for lunch. The correct answer is seven hours.

8. B (Standard Assessed: Algebra)
The only answer choice that works is $y < 7$, since 6×7 is 42.

9. B (Standard Assessed: Data Analysis)
The denominator is 30, since this is how many marbles are in the bag. There are five red marbles in the bag. This fraction reduces to $\frac{1}{6}$.

10. C   **(Standard Assessed: Algebra)**

You can simplify the expression by canceling out the *y* in the numerator and one of the *y*'s in the denominator.

11. C   **(Standard Assessed: Data Analysis)**

There are 20 coins in the box and $\frac{3}{10}$ of these coins are nickels, but the fraction $\frac{3}{10}$ is reduced. It is $\frac{6}{20}$ before it is reduced. Therefore, 14 of the coins are pennies.

12. A   **(Standard Assessed: Measurement and Geometry)**

Point A matches the coordinates.

13. C   **(Standard Assessed: Measurement and Geometry)**

The only answer choice that could possibly be the temperature of a swimming pool in the summer is 85 degrees. The other answer choices are either too cold or too hot.

14. C   **(Standard Assessed: Numbers and Computation)**

To change a decimal to a percentage, move the decimal point two places to the right and add the percent sign.

15. B   **(Standard Assessed: Measurement and Geometry)**

There are 1,000 meters in one kilometer, so there are 2,000 meters in two kilometers.

16. B   **(Standard Assessed: Algebra)**

Answer choice B is the correct proportion, since we are unsure of the number of gallons of gasoline Lindsay needs to drive 200 miles.

17. C   **(Standard Assessed: Measurement and Geometry)**

If you multiply 12 by 6 and then the answer by 4, you get 288.

18. D   **(Standard Assessed: Data Analysis)**

There are three even numbers on a number cube and three odd numbers.

19. D   **(Standard Assessed: Algebra)**

Melanie worked 40 hours, for which she earned $8 an hour. She earned $320 for the 40 hours. Then she worked eight hours of overtime, for which she was paid $12 an hour. She earned $96 in overtime. If you add $320 + $96, you get $416.

20. B   (Standard Assessed: Measurement and Geometry)
Point D is located at the coordinates (4, 1). Four is along the *x*-axis and one is along the *y*-axis.

21. D   (Standard Assessed: Numbers and Computation)
According to the commutative property, 3×24 is the same as 24×3.

22. D   (Standard Assessed: Algebra)
If you substitute 4 for *y* and 6 for *x*, the expression looks like this: $2(4) + 10(6)$ or $8 + 60$. The correct answer is 68.

23. C   (Standard Assessed: Measurement and Geometry)
If you multiply 570 by 4, the answer is 2,280.

24. C   (Standard Assessed: Data Analysis)
To find out how many girls play basketball and soccer, you need to look in the portion of the basketball and soccer circles that does not include the portion in all three circles.

25. A   (Standard Assessed: Data Analysis)
For this problem, you need to look at the space where all three circles overlap.

26. D   (Standard Assessed: Algebra)
To solve this problem, you need to determine the total number of meets, which is 7, and the number of meets Alex ran, which is 5.

27. A   (Standard Assessed: Numbers and Computation)
The identity element for addition is zero.

28. C   (Standard Assessed: Algebra)
If you substitute 2 into the expression, it looks like this: $2(2)^3 + 4$. The correct answer is 20.

29. C   (Standard Assessed: Data Analysis)
When you add Beth's scores, you get 190. When you divide this by 10, the number of game scores, you get 19.

30. D (Standard Assessed: Numbers and Computation)
The additive inverse of -10 is 10.

31. C (Standard Assessed: Data Analysis)
The mode is the number that occurs most; in this case, it is 90.

32. C (Standard Assessed: Algebra)
To solve this problem, you have to substitute 14 for x: $2(14) - 3$. The correct answer is 25.

33. D (Standard Assessed: Numbers and Computation)
Calculating a company's taxes would be difficult. The best answer choice is D: computer.

34. C (Standard Assessed: Measurement and Geometry)
To answer this problem, you need to add 10 and 8 and subtract this number from 28.

35. C (Standard Assessed: Algebra)
For this problem, you need to substitute 3 into the expression, which gives you $2(3)^3 - 5$ or $2(27) - 5$. $54 - 5 = 49$.

36. B (Standard Assessed: Algebra)
To solve the equation $y + 9 = 12$, you need to subtract 9 from 12.

37. D (Standard Assessed: Numbers and Computation)
With scientific notation, you move the decimal over so that it is between the first and second numbers. The number of places you move the decimal is the power of 10.

38. D (Standard Assessed: Algebra)
Answer choice D is the only statement that can be represented by the algebraic expression $12 - y$, since Hakim has 12 stamps but gives y number of stamps to his brother.

39. B (Standard Assessed: Numbers and Computation)
An approximate number would be fine if you were counting the number of students in a school who play sports.

40. A (Standard Assessed: Data Analysis)
A bar graph is the best kind of graph to show the number of people in five towns who receive a morning newspaper. A circle graph shows parts of a whole, a Venn diagram

shows how things are alike and different, and a line graph shows trends related to two variables.

41. B (Standard Assessed: Measurement and Geometry)
To solve this problem, you can set up a proportion: $\frac{8}{10} = \frac{24}{x}$. The ratio between 8 and 24 is 1:3, so the correct answer is 30.

42. D (Standard Assessed: Algebra)
To solve this problem, you need to substitute 4 for x: $7(4)^2 + 10$ or $7(16) + 10$.

43. A (Standard Assessed: Measurement and Geometry)
Ounces is the most appropriate unit of measure for a small piece of cloth, because cloth is very light.

44. C (Standard Assessed: Number and Computation)
To estimate the sum of these numbers, round them to the nearest ten: 40, 10, 20, 20, and 20.

45. B (Standard Assessed: Data Analysis)
If you subtract the range from the age of the oldest person, you'll get the age of the youngest person, which is 11.

46. B (Standard Assessed: Algebra)
Since Malcolm is 17, you need to subtract 3 from this number to find Salma's age.

47. B (Standard Assessed: Measurement and Geometry)
You need a right angle to use the Pythagorean theorem; answer choice B is the only triangle with a right angle.

48. D (Standard Assessed: Numbers and Computation)
To solve this problem, multiply $8 by 12, the number of hours Madeline works per week. Then multiply this number by 10, the number of weeks in the summer. You get 960, which approximates answer choice D and is thus the best answer.

49. A
The coordinates of point A are $(4, -2)$.

50. C (Standard Assessed: Measurement and Geometry)
The flag is simply slid in one direction, so the transformation is a translation.

51. A (Standard Assessed: Measurement and Geometry)
The mass of a box of cereal is best measured in grams. The other units of measurement would be too heavy.

52. A (Standard Assessed: Algebra)
The expression "x decreased by 2 is equal to $\frac{1}{3}$ of y" is $x - 2 = \frac{y}{3}$.

53. D (Standard Assessed: Algebra)
If you plug the correct amounts into the formula, it looks like this: $V = 3.14 \times 9 \times 12$.

54. D (Standard Assessed: Measurement and Geometry)
These angles are supplementary; they add up to 180 degrees. To find the answer, subtract the value of the given angle, 45 degrees, from 180.

55. C (Standard Assessed: Algebra)
To solve this problem, substitute 10 for x: $3(10) - 2 = 28$. Then subtract the number of girls, 10, to get the number of boys. $28 - 10 = 18$.

56. A (Standard Assessed: Measurement and Geometry)
Point A has the coordinates (−1, 1).

57. C (Standard Assessed: Numbers and Computation)
To find the discount, multiply $75 by .20.

58. D (Standard Assessed: Algebra)
To find the value of this equation, put y on one side by itself: $y = \frac{(16 + 4)}{2}$. The correct answer is 10.

59. C (Standard Assessed: Algebra)
If you substitute the radius and height of the cylinder into the formula, it looks like this: $V = 3.14 \times 16 \times 8$. The answer is approximately 402.

60. A (Standard Assessed: Measurement and Geometry)
Moving a geometric figure in one direction without turning it is called a translation.

Index

A
Accommodations, 2–3
Acute angles, 132
Addition
    associative property, 23–24
    commutative property, 22–23
Adjacent angles, 132
Algebraic equations, 148–149
Algebraic expressions, 143–147, 149
Angles, 131, 132–133, 137
Area, 96–97
Associative property, 23–24

B
Bar graphs, 76–77

C
Calculator use
    discounts, 45
    powers, 17
    scientific notation, 19
    square roots, 20
Capacity, 90, 93
Circle graphs, 78
Circles
    area, 96–97
    central angles, 137
    circumference, 126, 128
    description, 127
Circumference, 126, 128
Clustering, 39
Commutative property, 22–23
Comparisons, 14–15
Complementary angles, 133
Computational tools, 26–27
Congruent figures, 108
Conversions
    decimal into percent, 16
    percent into decimal, 15–16
    percent into fraction, 16
Coordinate planes, 113–115
Cube of number, 17
Customary units, 89–90
Cylinders, 97

D
Decimals, 14–15, 16
Denominators, 12
Disabled students, 2–3
Discounts, 44–46
Distributive property, 144

E
Endpoints, 131
Equations, 148–149, 163–164
Equilateral triangles, 135
Equivalent fractions, 12–13
Equivalent numbers, 12
Estimation, 38–42
Exponents, 16–17
Expressions, 143–147, 149

F
Formulas, 149
Fractions, 12–13, 16
Front-end estimation, 39–40

G
Georgia High School Graduation Test (GHSGT)
    about, 1–2
    accommodations, 2–3
    overview, 4–5
    standards, 5–8
    study tips, 4
    test-taking tips, 9–10
    when/where given, 2
Graphs, 75–78

H
Hexagons, 127
Hypotenuse, 135

I
Identity elements, 24–25
Inequalities, 26, 159, 162–163
Interest, 46–47
Inverse operation, 20, 25
Isosceles triangles, 135

L
Legs, 135
Length, 90, 93
Linear equations, 163–164
Linear inequalities, 162–163
Line graphs, 77
Lines, 131, 167
Line segments, 131

M
Mass, 90, 93
Mean, 61–62
Measurement
    estimations, 40–41
    systems of, 89–90
Measures of central tendency, 55
Median, 62
Metric system, 92–93
Mixed numbers, 13
Mode, 62
Money problems, 44–47
Multiplication
    associative property, 23–24
    commutative property, 22–23
    of fractions, 13

N
Numerators, 12

O
Obtuse angles, 132
Octagons, 127
Opposites, 20, 25

P
Parallel lines, 131
Parallelograms, 126
Patterns, 22
Pentagons, 127
Percents, 15–16, 41
Perimeter, 126–128
Perpendicular lines, 131
Pi (π), 128
Pictographs, 75–76
Pie charts or graphs, 78
Plane figures, 126–128
Point of intersection, 131
Powers, 16–17
Principal, 46
Probability, 55–58
Properties, 22–26
Proportions, 160–161
Pythagorean theorem, 135–136

R
Radius, 128
Range, 62–63
Rate problems, 149
Rays, 131
Reciprocals, 25
Rectangles
   area, 96
   description, 126
   similar, 109
Rectangular solids, 97
Reducing fractions, 16
Reference points, 40
Reflection, 112
Reflex angles, 132
Repeating decimals, 15
Rhombuses, 126
Right angles, 132
Right triangles, 135
Rise over run formula, 167
Rotation, 111
Rounding, 38–39

S
Sale prices, 44–46
Scalene triangles, 135

Scientific notation, 17–19
Similar figures, 108–109
Situations, 41–42
Slope, 167
Square of number, 17
Square roots, 20
Squares, 126
Standards, 5–8
Straight angles, 132
Supplementary angles, 133

T
Temperature, 90
Terminating decimals, 15
Time, 95
Transformations, 111–112
Translation, 112
Transversals, 133
Trapezoids, 126
Tree diagrams, 57
Triangles
   congruent, 108
   description, 126, 134
   Pythagorean theorem, 135–136
   similar, 108
   types of, 135

U
Undefined slope, 167

V
Variables, 143
Venn diagrams, 78–79
Vertex (vertices)
   central angles, 137
   definition, 131
   triangles, 134
Vertical angles, 133
Volume, 97

W
Waivers, 2–3
Weight, 90, 93

X
x-axis, 113–114
x-intercept, 167

Y
y-axis, 113–114
y-intercept, 167